HEATING
AND AIR
CONDITIONING

THE ILLUSTRATED HOME SERIES

ALSO AVAILABLE IN THE ILLUSTRATED HOME SERIES

STRUCTURE, ROOFING AND THE EXTERIOR

ELECTRICAL, PLUMBING, INSULATION AND THE INTERIOR

ALSO BY ALAN CARSON AND ROBERT DUNLOP

INSPECTING A HOUSE: A GUIDE FOR BUYERS, OWNERS, AND RENOVATORS

HEATING
AND AIR
CONDITIONING

THE ILLUSTRATED HOME SERIES

Carson Dunlop & Associates

Stoddart

Published in 2000 by Stoddart Publishing Co. Limited
34 Lesmill Road, Toronto, Canada M3B 2T6
180 Varick Street, 9th Floor, New York, New York 10014

Distributed in Canada by:
General Distribution Services Ltd.
325 Humber College Blvd., Toronto, Ontario M9W 7C3
Tel. (416) 213-1919   Fax (416) 213-1917
Email cservice@genpub.com

Distributed in the United States by:
General Distribution Services Inc.
PMB 128, 4500 Witmer Industrial Estates, Niagara Falls, New York 14305-1386
Toll-free Tel.1-800-805-1083   Toll-free Fax 1-800-481-6207
Email gdsinc@genpub.com

04 03 02 01 00 1 2 3 4 5

Canadian Cataloguing in Publication Data

Carson, Alan
Heating and air conditioning
(Illustrated home series)
ISBN 0-7737-6146-2
1. Dwellings — Heating and ventilation — Pictorial works.
2. Dwellings — Heating and ventilation — Maintenance and repair — Pictorial works.
3. Dwellings — Air conditioning — Pictorial works. 4. Dwellings — Air conditioning — Maintenance and repair — Pictorial works.
I. Dunlop, Robert. II. Title. III. Series: Carson, Alan. Illustrated home series.
TH7222.C37 2000   697'.00028'8   C00-931478-4

U.S. Cataloguing in Publication Data
(Library of Congress Standards)

Carson, Alan.
Heating and air conditioning / Alan Carson and Robert Dunlop. — 1st ed.
[128]p.: ill. ; cm. (The Illustrated Home Series)
ISBN: 0-7737-61462 (pbk.)
1. Heating — Amateurs' manuals. 2. Air conditioning — Amateurs' manuals.
3. Dwellings — Heating and ventilation — Amateurs' manuals. 4. Dwellings — Maintenance and repair —
Amateurs' manuals. 1. Dunlop, Robert. II. Title. III. Series
697 21 2000 CIP

Cover Design: Tannice Goddard
Text Illustrations: VECTROgraphics
Text Design: Neglia Design Inc./Tannice Goddard

THE CANADA COUNCIL | LE CONSEIL DES ARTS
FOR THE ARTS | DU CANADA
SINCE 1957 | DEPUIS 1957

*We acknowledge for their financial support of our
publishing program the Canada Council, the Ontario Arts
Council, and the Government of Canada through the
Book Publishing Industry Development Program (BPIDP).*

Printed and bound in Canada

697
CAR   4/12/01   OCLC

CONTENTS

INTRODUCTION

At Carson Dunlop & Associates, a consulting engineering firm, our principle business is home inspection. A few years ago, we set out to build a distance education product for those planning to enter the home inspection business. We developed the Carson Dunlop Home Study System, the most comprehensive home inspection training program in existence.

Early in the development process, we were unable to find existing illustrations that would *show* our students what we were explaining to them in the text. Even when we could find good illustrations, they were limited to a specific topic, and similar illustrations for other subjects couldn't be found at all.

Enter Peter Yeates, an engineer at Carson Dunlop & Associates, and owner of a graphics company called VectroGraphics. He solved our problem by designing over 1,500 illustrations for us. Peter is one of a handful of people on the planet who combines the technical expertise, computer skills, and aesthetics sense needed to create these illustrations.

Not only have these illustrations proven to be an effective tool in training home inspectors, but they have turned out to be very useful during home inspections. They allow us to explain situations to our clients, the potential homeowner.

The comments from students, other home inspectors, and, most importantly, the home buying public, has led us to assemble the illustrations from the Home Study System and publish them as a series of books. Finally, you don't have to feel guilty just looking at the pictures.

HEATING AND AIR CONDITIONING

FURNACES

Most heating systems fall into two categories: furnaces and boilers. Furnaces use air as the heat transfer medium, while boilers use water. A furnace or a boiler can be fueled by gas, oil, electricity, wood, or almost any source of heat.

Generally, the heating illustrations are laid out component by component. We start with furnaces and deal with combustion air, the burner, the heat exchanger (the most critical component of a furnace), and the heating cabinet itself.

Next, some of the controls are illustrated, such as the fan limit switch and thermostat.

After the fuel has been burned, the products of combustion must be vented to the outdoors, so there are a series of illustrations showing various venting systems. At the same time, the heat extracted from the exhaust gases must be distributed through the house. There are several illustrations of ductwork.

The last furnace illustrations deal with mid- and high-efficiency (condensing) furnaces.

BOILERS

Hot water heating systems come next. Because it's a good possibility that a hot water system will over-heat, creating steam and an explosion, there are many safety controls on boilers. Most of these are illustrated.

Once the water is hot, it must be distributed through the house. Several illustrations showing various water distribution systems are included.

MISCELLANEOUS

Chimneys are the next order of business. All of the products of combustion must be exhausted out of the building. There are several illustrations here that show how this can be done.

The less common heating systems are dealt with in the following order: wood heating systems, steam systems, and electric systems. Finally, permanently installed space heaters are illustrated.

AIR CONDITIONING

Few people truly understand air conditioning systems (heat pumps are even less understood). The illustrations in this section will show you the components that are buried inside the ductwork or the outdoor unit. These pictures should demystify air conditioning and heat pump systems.

INSTRUCTIONS FOR USE

You'll notice that there's a table of contents at the beginning of each part of this book. The heading of each illustration is listed next to the number of the illustration. Simply flip through the book until you find the illustration you want to look at.

PART 1

HEATING

FURNACES — GAS & OIL

GENERAL

HEAT TRANSFER

GAS PIPING AND METERS

COMBUSTION AIR

GAS BURNERS

HEAT EXCHANGERS

FURNACE CABINETS

FAN/LIMIT SWITCHES

THERMOSTATS

VENTING GAS FURNACES

HOT WATER BOILERS

GENERAL

0128 How boilers work
0129 How radiators heat the air through convection
0130 Radiator covers
0131 Superheated water

HEAT EXCHANGERS

0132 Heat exchangers
0133 Copper tube heat exchangers

CONTROLS

0134 Pressure relief valve
0135 Pressure relief valve location
0136 Inspecting pressure relief valves
0137 High temperature limit switch
0138 Low water cut out
0139 Backflow preventer
0140 Pressure reducing valve location
0141 Backflow preventer installed backwards
0142 Temperature and pressure gauge
0143 Pressure reducing valve
0144 Combined pressure reducing valve and pressure relief valve
0145 Pressure set too low
0146 Don't operate air bleed valves
0147 Pressure reducing/relief valve installed backwards
0148 Air separators
0149 Aquastat – primary control
0150 Inoperative aquastat
0151 Pump control
0152 Zone control with pumps
0153 Zone control with valves
0154 Outdoor air temperature sensor
0155 Flow control valves

DISTRIBUTION SYSTEMS

0156 Open hydronic system
0157 Closed hydronic system
0158 Series loop
0159 One-pipe system
0160 Two-pipe system (direct return)
0161 Two-pipe system (reverse return)
0162 Balancing methods
0163 Expansion tank water levels
0164 Overflow pipe
0165 Conventional expansion tank
0166 Diaphragm tank
0167 Circulating pump
0168 Pipe corrosion
0169 Extending hot water systems
0170 Covering radiators reduces efficiency
0171 Convector
0172 Finned tube baseboard
0173 Cast iron baseboard
0174 Mixing systems
0175 Radiators on ceilings or high on walls
0176 Radiator valve leaks
0177 Hot water radiant heat
0178 Water blender on radiant system
0179 Tankless coil
0180 Side arm heater
0181 Spillage switch
0182 Bypass loop

PULSE COMBUSTION

0183 Pulse combustion – how it works

CHIMNEYS

GENERAL

0184 Chimneys
0185 Vents
0186 Chimneys are not supporting structures
0187 Warm chimneys are best
0188 Chimney extender
0189 Masonry chimneys

MASONRY CHIMNEYS

0190 Basic masonry chimney components
0191 Basic masonry chimney and fireplace components
0192 Chimney walls
0193 Lateral support for masonry chimneys
0194 Clay tile flue liners

WOOD HEATING SYSTEMS

GENERAL

FURNACES AND BOILERS

WOOD STOVES (SPACE HEATERS)

How heat is generated

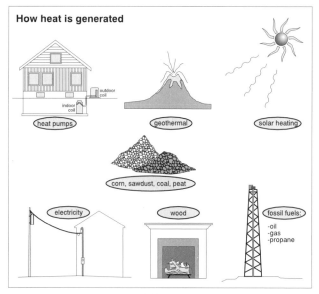

outdoor coil

indoor coil

heat pumps

geothermal

solar heating

corn, sawdust, coal, peat

electricity

wood

fossil fuels:
-oil
-gas
-propane

0 0 0 1

British Thermal Unit (BTUs)

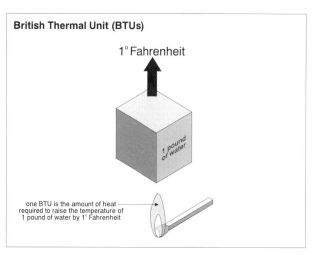

1° Fahrenheit

1 pound of water

one BTU is the amount of heat required to raise the temperature of 1 pound of water by 1° Fahrenheit

0 0 0 2

Cool spots

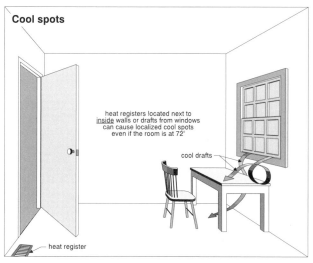

heat registers located next to <u>inside</u> walls or drafts from windows can cause localized cool spots even if the room is at 72°

cool drafts

heat register

0 0 0 3

Cold floors

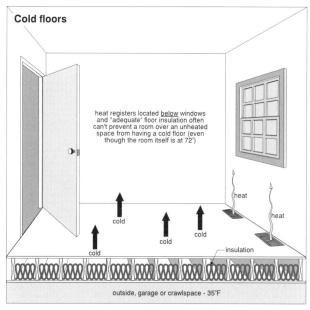

heat registers located <u>below</u> windows and "adequate" floor insulation often can't prevent a room over an unheated space from having a cold floor (even though the room itself is at 72°)

heat

heat

cold

cold

cold

cold

insulation

outside, garage or crawlspace - 35°F

0 0 0 4

The outdoor fire

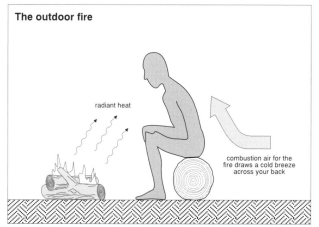

radiant heat

combustion air for the fire draws a cold breeze across your back

0 0 0 5

Fire comes inside

radiant heat

radiant heat

radiant heat bounces back off the cave walls - keeping your back warmer

unfortunately smoke is a problem

0 0 0 6

The hole in the roof

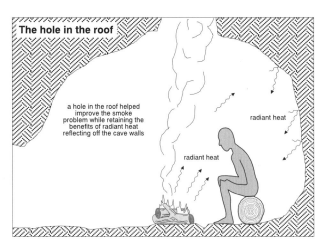

a hole in the roof helped improve the smoke problem while retaining the benefits of radiant heat reflecting off the cave walls

radiant heat

radiant heat

0 0 0 7

Homes built with holes

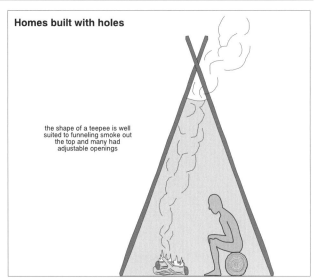

the shape of a teepee is well suited to funneling smoke out the top and many had adjustable openings

0 0 0 8

Fire gets contained

people started to contain fire in "boxes" of stone or steel

they found that even though they couldn't "see" the fire they were still warmed by the walls of the box

smoke exits through hole in roof

radiant heat

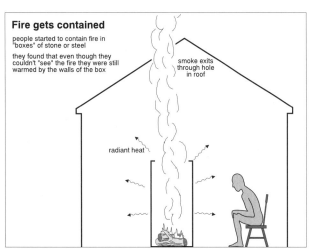

0 0 0 9

Building a chimney

chimney

hinged door for adding wood and letting air in for combustion

radiant heat

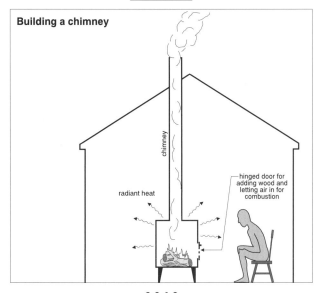

0 0 1 0

Gravity furnace

heat registers are on inside walls

second floor

usually only one cold air return (on the main floor) and it is often located near an outside wall

several registers are often fed off one duct

first floor

air return

basement

supply duct

supply ducts have exaggerated upward slope and are large and round

furnace is centrally located

cross section

no fan and no filter

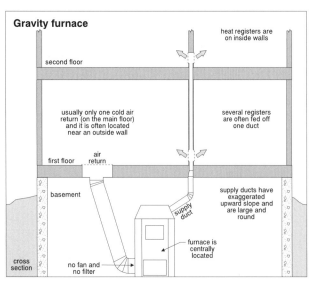

0 0 1 1

House air flow

air flow

cool air falls

warm air rises

return grill

heat register

cold air 70°F

air **pulled** through return ducts

air **pushed** through supply ducts

hot air 140°F

air flow

air flow

heat exchanger

air filters

blower motor

burner

blower

cross section

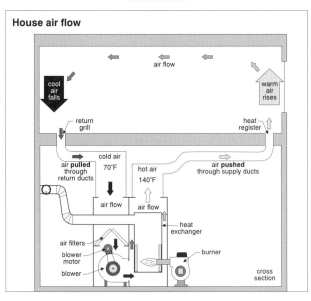

0 0 1 2

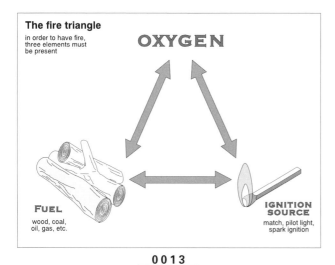

The fire triangle

in order to have fire, three elements must be present

OXYGEN

FUEL
wood, coal, oil, gas, etc.

IGNITION SOURCE
match, pilot light, spark ignition

0013

Methods of heat transfer

air above the pot warms up (becomes less dense) and rises - drawing more cool air in from the sides to be heated up

outside (cold)   inside (warm)

heat

radiation
heat transferred through electromagnetic waves e.g. thermal infrared energy (sunlight)

convection
heat transfer within a gas or liquid

conduction
heat transfer through a solid material

0014

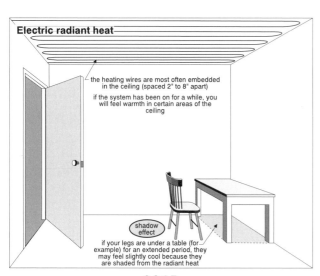

Electric radiant heat

the heating wires are most often embedded in the ceiling (spaced 2" to 8" apart)

if the system has been on for a while, you will feel warmth in certain areas of the ceiling

shadow effect

if your legs are under a table (for example) for an extended period, they may feel slightly cool because they are shaded from the radiant heat

0015

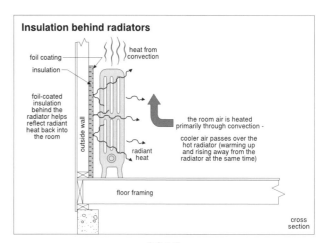

Insulation behind radiators

foil coating

insulation

heat from convection

foil-coated insulation behind the radiator helps reflect radiant heat back into the room

outside wall

radiant heat

the room air is heated primarily through convection -

cooler air passes over the hot radiator (warming up and rising away from the radiator at the same time)

floor framing

cross section

0016

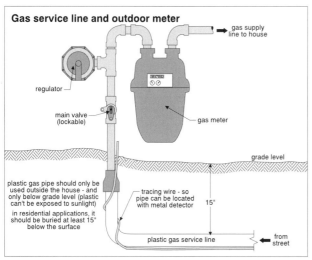

Gas service line and outdoor meter

gas supply line to house

regulator

main valve (lockable)

gas meter

grade level

plastic gas pipe should only be used outside the house - and only below grade level (plastic can't be exposed to sunlight)

in residential applications, it should be buried at least 15" below the surface

tracing wire - so pipe can be located with metal detector

15"

plastic gas service line

from street

0017

Ice on regulator

ice build-up on regulators can block the vents and potentially allow excess gas pressure into the house

this is most likely to happen when the meter is below the drip line of the roof

gas supply line to house

regulator

ice

gas meter

snow

grade level

0018

Poor meter locations

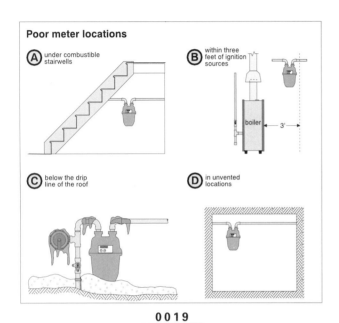

(A) under combustible stairwells

(B) within three feet of ignition sources

boiler   3'

(C) below the drip line of the roof

(D) in unvented locations

0019

Gas piping terminology

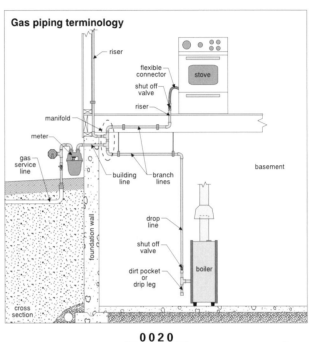

riser

stove

flexible connector

shut off valve

riser

manifold

meter

gas service line

building line

branch lines

basement

drop line

shut off valve

dirt pocket or drip leg

boiler

foundation wall

cross section

0020

Flexible gas appliance connectors

flexible connectors are used to connect semi-portable appliances such as gas ranges, dryers and outdoor barbecues to the gas supply piping

the illustration shows some examples of improper installation

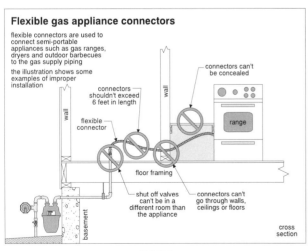

connectors can't be concealed

connectors shouldn't exceed 6 feet in length

wall

flexible connector

wall

range

floor framing

shut off valves can't be in a different room than the appliance

connectors can't go through walls, ceilings or floors

basement

cross section

0021

Teflon tape at connections

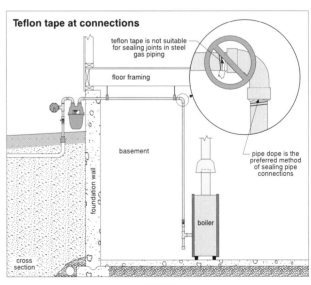

teflon tape is not suitable for sealing joints in steel gas piping

floor framing

basement

pipe dope is the preferred method of sealing pipe connections

foundation wall

boiler

cross section

0022

Flexible copper tubing for gas piping

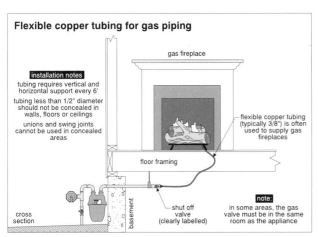

gas fireplace

installation notes

tubing requires vertical and horizontal support every 6'

tubing less than 1/2" diameter should not be concealed in walls, floors or ceilings

unions and swing joints cannot be used in concealed areas

flexible copper tubing (typically 3/8") is often used to supply gas fireplaces

floor framing

shut off valve (clearly labelled)

note:
in some areas, the gas valve must be in the same room as the appliance

cross section

basement

0023

Gas shut off valves

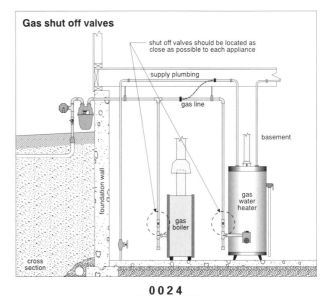

shut off valves should be located as close as possible to each appliance

supply plumbing

gas line

basement

foundation wall

gas boiler

gas water heater

cross section

0024

Grounding the gas piping

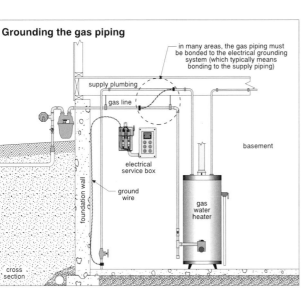

in many areas, the gas piping must be bonded to the electrical grounding system (which typically means bonding to the supply piping)

supply plumbing

gas line

basement

electrical service box

ground wire

foundation wall

gas water heater

cross section

0025

Gas piping support

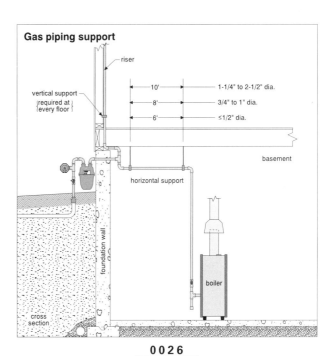

riser

vertical support (required at every floor)

10'	1-1/4" to 2-1/2" dia.
8'	3/4" to 1" dia.
6'	≤1/2" dia.

basement

foundation wall

horizontal support

boiler

cross section

0026

Drip leg

the drip leg (or dirt pocket) serves as a collection area for sediment to reduce the chance of clogged gas valves or burners

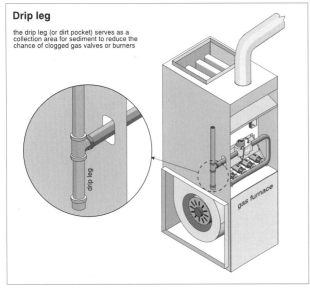

drip leg

gas furnace

0027

Combustion air

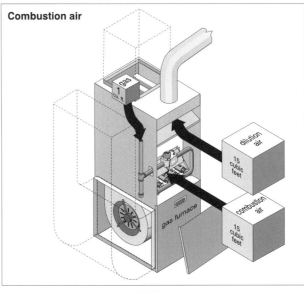

gas 1 cu. ft.

dilution air 15 cubic feet

combustion air 15 cubic feet

gas furnace

0028

Three kinds of draft

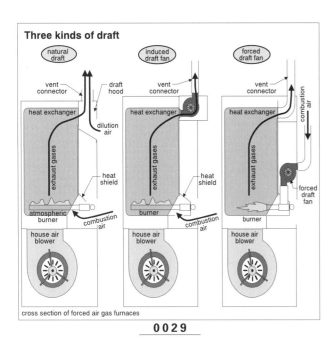

natural draft

induced draft fan

forced draft fan

vent connector

draft hood

heat exchanger

dilution air

exhaust gases

heat shield

atmospheric burner

combustion air

house air blower

vent connector

heat exchanger

exhaust gases

heat shield

burner

combustion air

house air blower

vent connector

combustion air

heat exchanger

exhaust gases

burner

forced draft fan

house air blower

cross section of forced air gas furnaces

0029

Natural draft burners

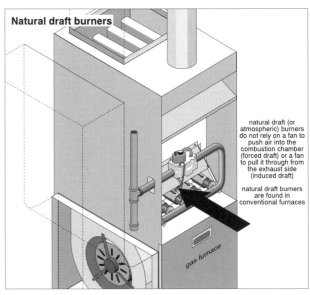

natural draft (or atmospheric) burners do not rely on a fan to push air into the combustion chamber (forced draft) or a fan to pull it through from the exhaust side (induced draft)

natural draft burners are found in conventional furnaces

gas furnace

0030

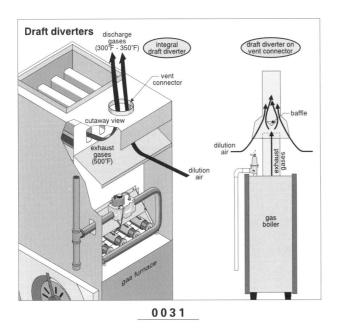

Draft diverters

discharge gases (300°F - 350°F)

integral draft diverter

vent connector

cutaway view

exhaust gases (500°F)

dilution air

gas furnace

draft diverter on vent connector

baffle

dilution air

exhaust gases

gas boiler

0031

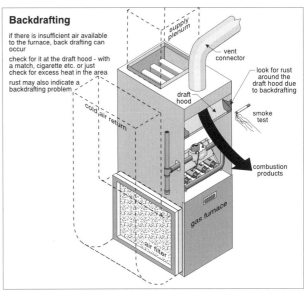

Backdrafting

if there is insufficient air available to the furnace, back drafting can occur

check for it at the draft hood - with a match, cigarette etc. or just check for excess heat in the area

rust may also indicate a backdrafting problem

supply plenum

vent connector

look for rust around the draft hood due to backdrafting

draft hood

smoke test

cold air return

combustion products

air filter

gas furnace

0032

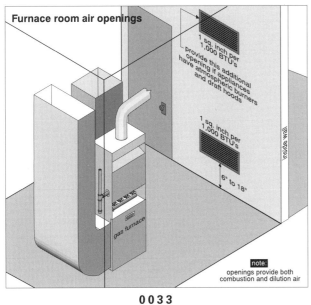

Furnace room air openings

1 sq. inch per 1,000 BTU's
provide this additional opening if appliances have atmospheric burners and draft hoods

1 sq. inch per 1,000 BTU's

6" to 18"

inside wall

gas furnace

note:
openings provide both combustion and dilution air

0033

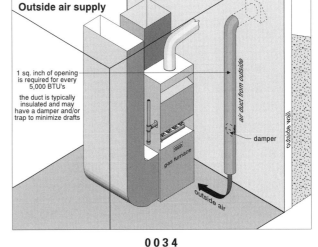

Outside air supply

1 sq. inch of opening is required for every 5,000 BTU's

the duct is typically insulated and may have a damper and/or trap to minimize drafts

air duct from outside

damper

outside wall

gas furnace

outside air

0034

Gas burners

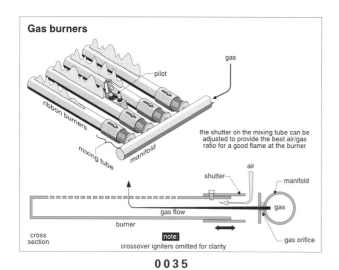

gas

pilot

ribbon burners

mixing tube

manifold

the shutter on the mixing tube can be adjusted to provide the best air/gas ratio for a good flame at the burner

shutter

air

manifold

gas flow

gas

burner

gas orifice

cross section

note: crossover igniters omitted for clarity

0035

Monoport burners

monoport burners are often found on newer gas furnaces and conversion gas burners

they are typically fan assisted

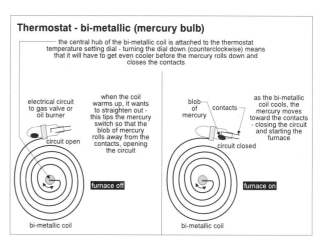

vent connector

heat exchanger

combustion air

modern gas furnace

exhaust gases

pilot

mono port burner

gas

manifold

burner

forced draft fan

perspective

cross section

0036

Gas supply to burners

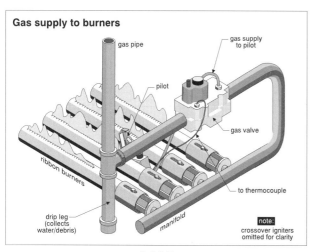

gas pipe

gas supply to pilot

pilot

gas valve

ribbon burners

to thermocouple

drip leg (collects water/debris)

manifold

note: crossover igniters omitted for clarity

0037

Crossover igniters

the crossover igniter is used to bridge the ignition flame from one burner to the next

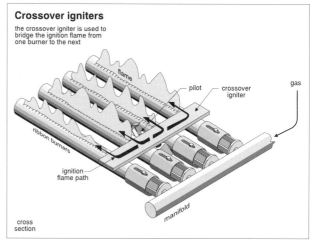

flame

pilot

crossover igniter

gas

ribbon burners

ignition flame path

manifold

cross section

0038

Refractory/fire pot

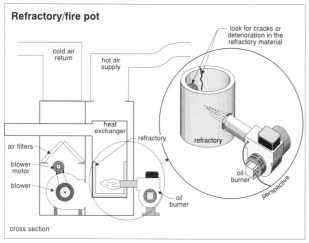

cold air return

hot air supply

look for cracks or deterioration in the refractory material

heat exchanger

refractory

air filters

refractory

blower motor

blower

oil burner

oil burner

perspective

cross section

0039

Thermostat - bi-metallic (mercury bulb)

the central hub of the bi-metallic coil is attached to the thermostat temperature setting dial - turning the dial down (counterclockwise) means that it will have to get even cooler before the mercury rolls down and closes the contacts

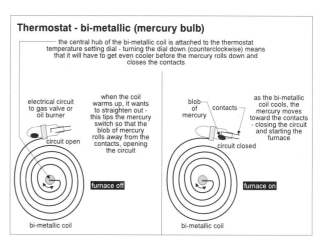

electrical circuit to gas valve or oil burner

when the coil warms up, it wants to straighten out - this tips the mercury switch so that the blob of mercury rolls away from the contacts, opening the circuit

circuit open

furnace off

bi-metallic coil

blob of mercury

contacts

as the bi-metallic coil cools, the mercury moves toward the contacts - closing the circuit and starting the furnace

circuit closed

furnace on

bi-metallic coil

0040

Pilot light - relighting

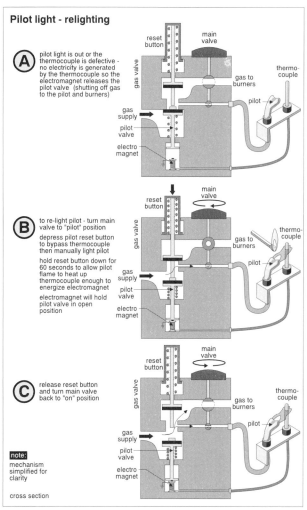

Ⓐ pilot light is out or the thermocouple is defective - no electricity is generated by the thermocouple so the electromagnet releases the pilot valve (shutting off gas to the pilot and burners)

reset button
main valve
gas valve
gas to burners
thermo-couple
gas supply
pilot
pilot valve
electro magnet

Ⓑ to re-light pilot - turn main valve to "pilot" position

depress pilot reset button to bypass thermocouple then manually light pilot

hold reset button down for 60 seconds to allow pilot flame to heat up thermocouple enough to energize electromagnet

electromagnet will hold pilot valve in open position

Ⓒ release reset button and turn main valve back to "on" position

note:
mechanism simplified for clarity

cross section

0042

Continuous pilot light

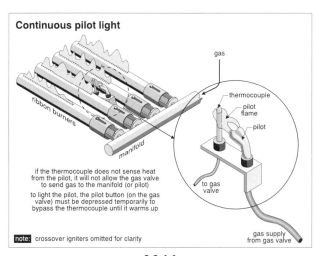

ribbon burners
manifold

gas
thermocouple
pilot flame
pilot

to gas valve

gas supply from gas valve

if the thermocouple does not sense heat from the pilot, it will allow the gas valve to send gas to the manifold (or pilot)

to light the pilot, the pilot button (on the gas valve) must be depressed temporarily to bypass the thermocouple until it warms up

note: crossover igniters omitted for clarity

0041

Scorching

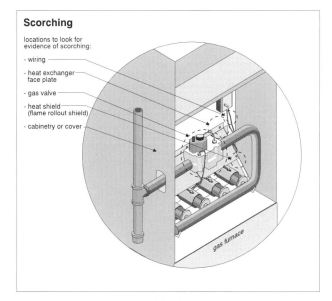

locations to look for evidence of scorching:

- wiring
- heat exchanger face plate
- gas valve
- heat shield (flame rollout shield)
- cabinetry or cover

gas furnace

0043

Heat shield or flame rollout shield

in forced air gas furnace

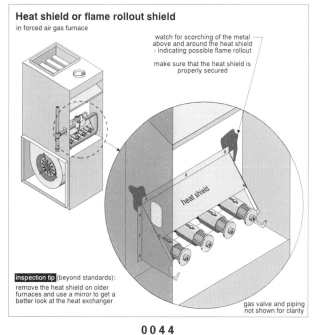

watch for scorching of the metal above and around the heat shield - indicating possible flame rollout

make sure that the heat shield is properly secured

heat shield

inspection tip (beyond standards):

remove the heat shield on older furnaces and use a mirror to get a better look at the heat exchanger

gas valve and piping not shown for clarity

0044

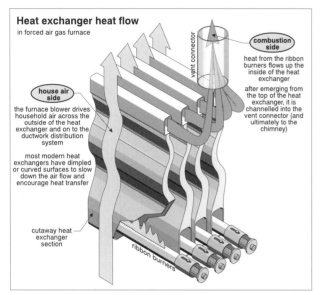

Heat exchanger heat flow
in forced air gas furnace

vent connector

combustion side

heat from the ribbon burners flows up the inside of the heat exchanger

after emerging from the top of the heat exchanger, it is channelled into the vent connector (and ultimately to the chimney)

house air side

the furnace blower drives household air across the outside of the heat exchanger and on to the ductwork distribution system

most modern heat exchangers have dimpled or curved surfaces to slow down the air flow and encourage heat transfer

cutaway heat exchanger section

ribbon burners

0045

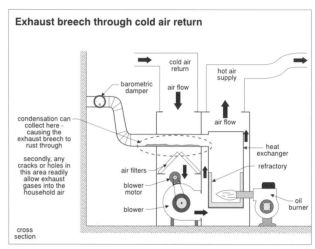

Exhaust breech through cold air return

cold air return

hot air supply

air flow

barometric damper

air flow

condensation can collect here - causing the exhaust breech to rust through

secondly, any cracks or holes in this area readily allow exhaust gases into the household air

air filters

blower motor

blower

heat exchanger

refractory

oil burner

cross section

0046

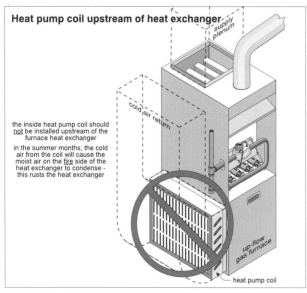

Heat pump coil upstream of heat exchanger

supply plenum

cold air return

the inside heat pump coil should <u>not</u> be installed upstream of the furnace heat exchanger

in the summer months, the cold air from the coil will cause the moist air on the <u>fire</u> side of the heat exchanger to condense - this rusts the heat exchanger

up-flow gas furnace

heat pump coil

0047

Downflow and upflow furnaces

return

vent connector

air filters

blower

draft hood

heat exchanger

gas valve

gas burners

air flow

supply

downflow gas furnace often used above short crawlspaces where access is limited

supply

return

vent connector

air flow

draft hood

heat exchanger

gas valve

gas burners

air filter

blower

upflow gas furnace most common - typically located in basement

0048

Horizontal gas-fired furnace

often found in crawlspaces or attics
(wherever headroom is limited)

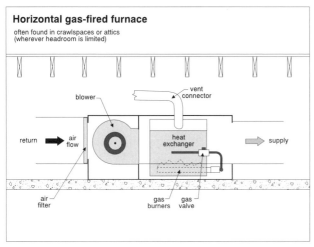

blower

vent
connector

return

air
flow

heat
exchanger

supply

air
filter

gas
burners

gas
valve

0049

Fan cover missing

operating a furnace without the fan
cover can be dangerous as
negative pressure can be created
in the furnace room - sucking
combustion products into the fan
compartment and blowing them
throughout the house

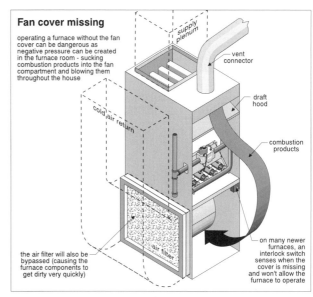

supply
plenum

vent
connector

draft
hood

cold air return

combustion
products

the air filter will also be
bypassed (causing the
furnace components to
get dirty very quickly)

air filter

on many newer
furnaces, an
interlock switch
senses when the
cover is missing
and won't allow the
furnace to operate

0050

Furnace covers

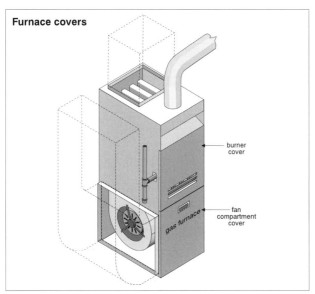

burner
cover

fan
compartment
cover

gas furnace

0051

Furnace clearances

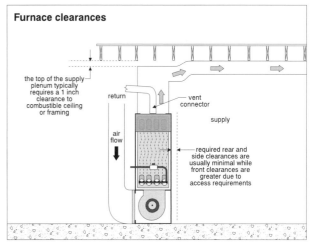

the top of the supply
plenum typically
requires a 1 inch
clearance to
combustible ceiling
or framing

return

air
flow

vent
connector

supply

required rear and
side clearances are
usually minimal while
front clearances are
greater due to
access requirements

0052

Fan switch

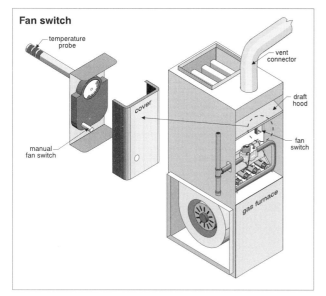

temperature probe

vent connector

cover

draft hood

manual fan switch

fan switch

gas furnace

0 0 5 3

Fan/limit switch

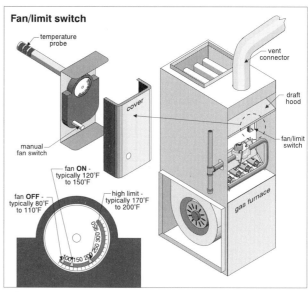

temperature probe

vent connector

cover

draft hood

manual fan switch

fan/limit switch

fan ON - typically 120°F to 150°F

fan OFF - typically 80°F to 110°F

high limit - typically 170°F to 200°F

gas furnace

0 0 5 4

Fan/limit switch set wrong or defective

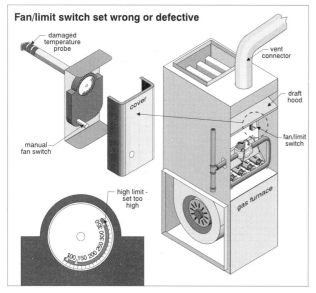

damaged temperature probe

vent connector

cover

draft hood

manual fan switch

fan/limit switch

high limit - set too high

gas furnace

0 0 5 5

Thermostat - bi-metallic (snap-action)

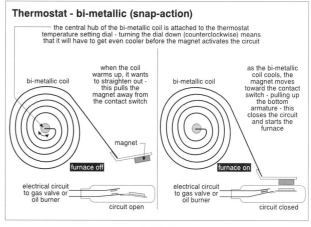

the central hub of the bi-metallic coil is attached to the thermostat temperature setting dial - turning the dial down (counterclockwise) means that it will have to get even cooler before the magnet activates the circuit

bi-metallic coil

when the coil warms up, it wants to straighten out - this pulls the magnet away from the contact switch

magnet

furnace off

electrical circuit to gas valve or oil burner

circuit open

bi-metallic coil

as the bi-metallic coil cools, the magnet moves toward the contact switch - pulling up the bottom armature - this closes the circuit and starts the furnace

furnace on

electrical circuit to gas valve or oil burner

circuit closed

0 0 5 6

Variable thermostat anticipator

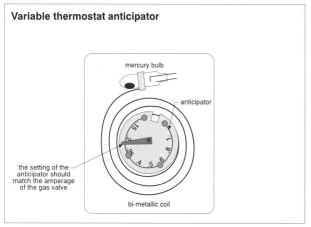

mercury bulb

anticipator

the setting of the anticipator should match the amperage of the gas valve

bi-metallic coil

0 0 5 7

Thermostat must be level

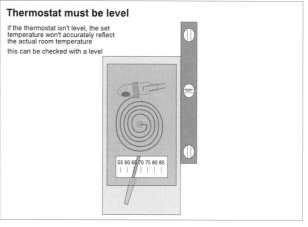

if the thermostat isn't level, the set temperature won't accurately reflect the actual room temperature

this can be checked with a level

55 60 65 70 75 80 85

0 0 5 8

Vent connector slope

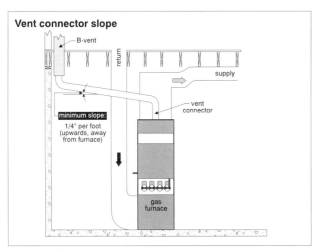

B-vent

return

supply

vent connector

minimum slope:
1/4" per foot
(upwards, away
from furnace)

gas
furnace

0059

Vent connector length

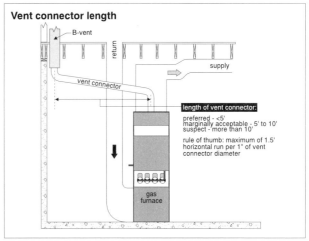

B-vent

return

supply

vent connector

length of vent connector:
preferred - <5'
marginally acceptable - 5' to 10'
suspect - more than 10'

rule of thumb: maximum of 1.5'
horizontal run per 1" of vent
connector diameter

gas
furnace

0060

Vent connector support

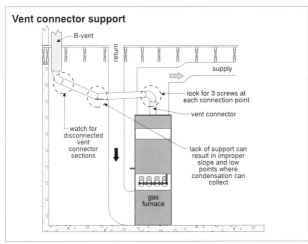

B-vent

return

supply

look for 3 screws at
each connection point

vent connector

watch for
disconnected
vent
connector
sections

lack of support can
result in improper
slope and low
points where
condensation can
collect

gas
furnace

0061

Combustible clearances for vent connectors and B-vents

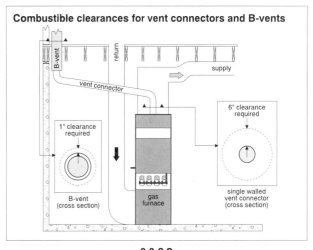

B-vent

return

supply

vent connector

1" clearance
required

6" clearance
required

B-vent
(cross section)

gas
furnace

single walled
vent connector
(cross section)

0062

Size of vent connector

the vent connector diameter should match
the size of the flue collar

if the vent connector is too large or too
small, condensation or spillage could result

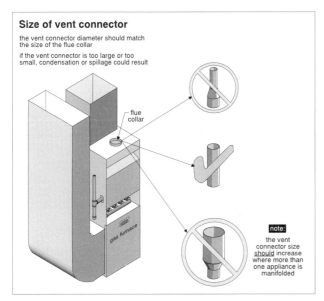

flue
collar

gas furnace

note:
the vent
connector size
should increase
where more
than one
appliance is
manifolded

0063

Chimney/vent connections

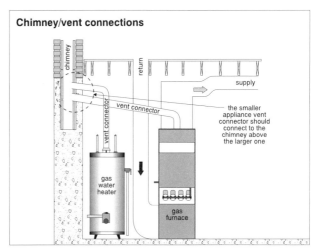

chimney

return

supply

vent connector

the smaller
appliance vent
connector should
connect to the
chimney above
the larger one

gas
water
heater

gas
furnace

0064

Vent connector extends too far into chimney

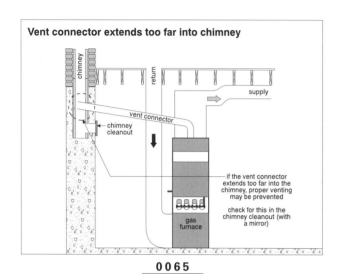

0065

Typical supply and return register locations

0066

Distribution system components for forced air heating

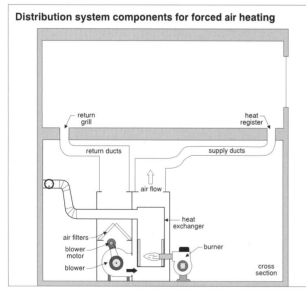

0067

Factors affecting air supply

1. duct size - bigger is better

2. duct shape -   round   oval   square   rectangular
   best shape ➡ worst shape

3. duct length - shorter is better

4. number of corners - fewer is better

5. duct type - flex duct can have pressure losses 3 times that of smooth duct

6. blower size - should be appropriate

7. blower speed - should be appropriate

8. blower blade profile

9. furnace location - central location is particularly important in larger homes

10. system extras - air filters and air conditioning coils can create significant pressure loss

0068

Blowers - belt drive and direct drive

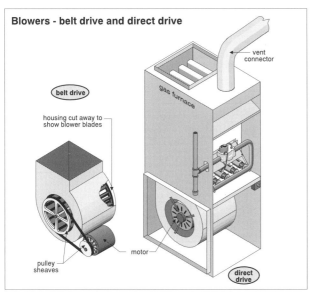

belt drive

housing cut away to show blower blades

gas furnace

vent connector

pulley sheaves

motor

direct drive

0069

Fan belts on furnace blowers

check belt for cracks or other wear
check belt tension (see below)
check for excess vibration
check for overheating at the motor

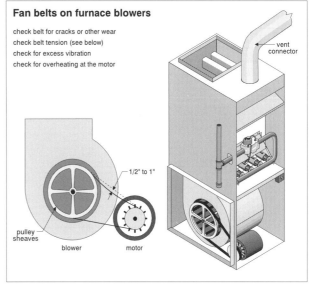

vent connector

1/2" to 1"

pulley sheaves

blower

motor

0070

Air filter orientation

typically, there is some sort of filter support on the <u>blower side</u> of the air filter

the filter must be installed with the "air flow" arrow properly oriented to make sure that the filter material doesn't get sucked into the blower

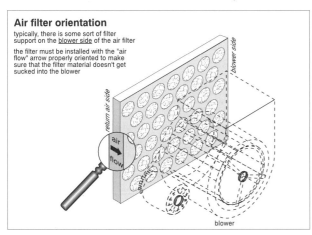

return air side

blower side

air flow

blower

0071

Electronic air cleaner

exploded view

charcoal filters - not always present

cells

supply plenum

return air side

air flow

blower side

cold air return

electronic air filter

gas furnace

prefilters

oppositely charged plates - attract and hold the dust

highly charged wires - charge the dust particles

test button

0072

Electronic air cleaner problems

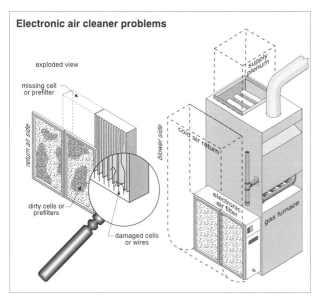

exploded view

missing cell
or prefilter

return air side

blower side

cold air return

supply plenum

dirty cells or
prefilters

damaged cells
or wires

electronic
air filter

gas furnace

0073

Electronic air cleaner - installed backwards/parts missing

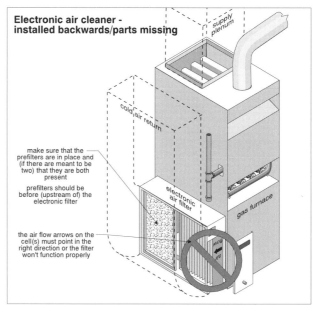

supply plenum

cold air return

make sure that the
prefilters are in place and
(if there are meant to be
two) that they are both
present

prefilters should be
before (upstream of) the
electronic filter

the air flow arrows on the
cell(s) must point in the
right direction or the filter
won't function properly

electronic
air filter

gas furnace

0074

Drum-type humidifier

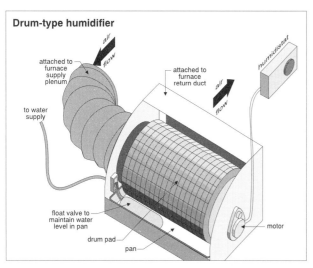

air flow

attached to
furnace
supply plenum

attached to
furnace
return duct

air flow

humidistat

to water
supply

float valve to
maintain water
level in pan

drum pad

pan

motor

0075

Trickle humidifier

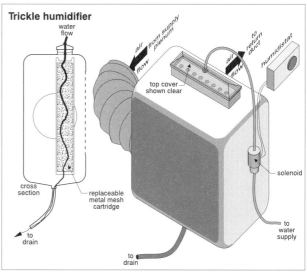

water flow

air from supply plenum

air flow

to return duct

humidistat

top cover
shown clear

cross section

replaceable
metal mesh
cartridge

to drain

to drain

solenoid

to water
supply

0076

Air flow through humidifier

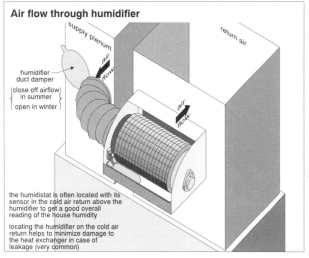

supply plenum

return air

air flow

humidifier
duct damper

(close off airflow)
in summer

open in winter

air flow

the humidistat is often located with its
sensor in the cold air return above the
humidifier to get a good overall
reading of the house humidity

locating the humidifier on the cold air
return helps to minimize damage to
the heat exchanger in case of
leakage (very common)

0077

Humidifier above heat exchanger

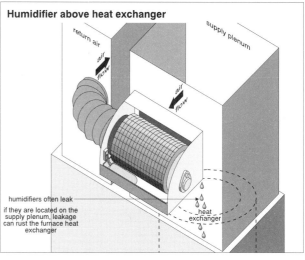

return air

supply plenum

air flow

air flow

humidifiers often leak

if they are located on the
supply plenum, leakage
can rust the furnace heat
exchanger

heat
exchanger

0078

Return ducts in floor joists

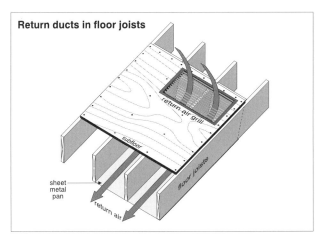

return air grill

subfloor

floor joists

sheet metal pan

return air

0079

Supply air registers, diffusers and grills

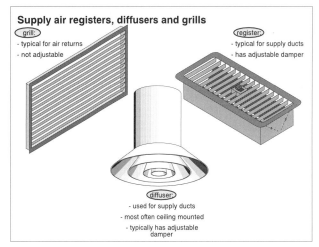

grill:
- typical for air returns
- not adjustable

register:
- typical for supply ducts
- has adjustable damper

diffuser:
- used for supply ducts
- most often ceiling mounted
- typically has adjustable damper

0080

Renovations remove ducts

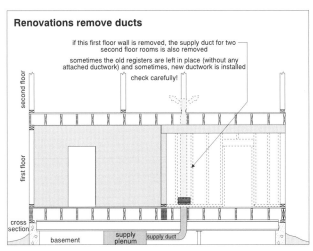

if this first floor wall is removed, the supply duct for two second floor rooms is also removed

sometimes the old registers are left in place (without any attached ductwork) and sometimes, new ductwork is installed

check carefully!

second floor

first floor

cross section

basement   supply plenum   supply duct

0081

High and low returns

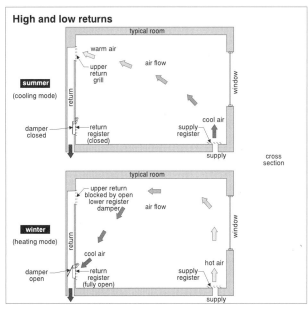

typical room

warm air

air flow

upper return grill

window

summer
(cooling mode)

return

damper closed

return register (closed)

supply register

cool air

supply

cross section

typical room

upper return blocked by open lower register damper

air flow

window

winter
(heating mode)

return

cool air

damper open

return register (fully open)

supply register

hot air

supply

0082

Ducts in concrete floor slabs

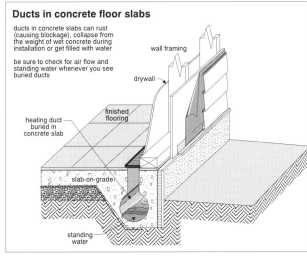

ducts in concrete slabs can rust (causing blockage), collapse from the weight of wet concrete during installation or get filled with water

be sure to check for air flow and standing water whenever you see buried ducts

wall framing

drywall

finished flooring

heating duct buried in concrete slab

slab-on-grade

standing water

0083

Supply and return registers poorly located

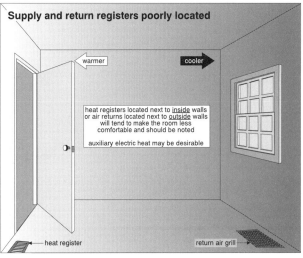

warmer

cooler

heat registers located next to inside walls or air returns located next to outside walls will tend to make the room less comfortable and should be noted

auxiliary electric heat may be desirable

heat register

return air grill

0084

Basement heat registers

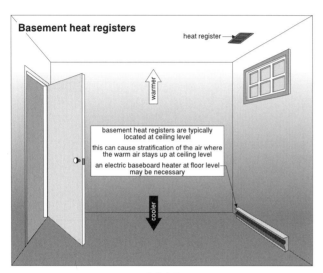

heat register

warmer

cooler

basement heat registers are typically located at ceiling level

this can cause stratification of the air where the warm air stays up at ceiling level

an electric baseboard heater at floor level may be necessary

0085

Return grill in furnace room

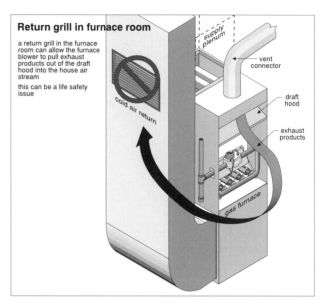

a return grill in the furnace room can allow the furnace blower to pull exhaust products out of the draft hood into the house air stream

this can be a life safety issue

supply plenum

cold air return

vent connector

draft hood

exhaust products

gas furnace

0086

Air return outside room

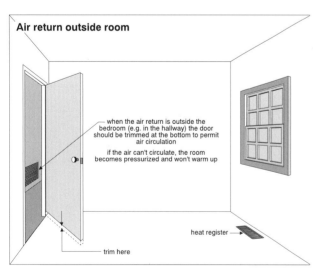

when the air return is outside the bedroom (e.g. in the hallway) the door should be trimmed at the bottom to permit air circulation

if the air can't circulate, the room becomes pressurized and won't warm up

trim here

heat register

0087

Intermittent pilot light

the igniter is very much like a spark plug - the spark jumping across the air gap lights the pilot flame

gas

ribbon burners

manifold

igniter

pilot

the flame sensor in an intermittent pilot system may be located near the pilot light or may be near one of the ribbon burners

to ignite power supply

gas supply from gas valve

note: crossover igniters omitted for clarity

0088

Hot surface ignition

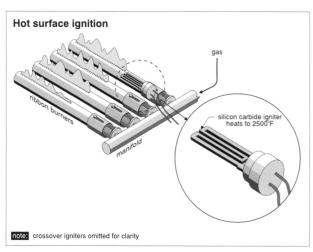

gas

ribbon burners

manifold

silicon carbide igniter
heats to 2500°F

note: crossover igniters omitted for clarity

0089

Blockage switches

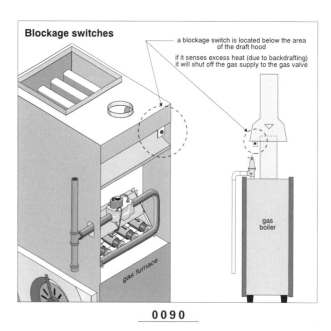

a blockage switch is located below the area
of the draft hood

if it senses excess heat (due to backdrafting)
it will shut off the gas supply to the gas valve

gas furnace

gas
boiler

0090

Improper side wall venting of gas appliances

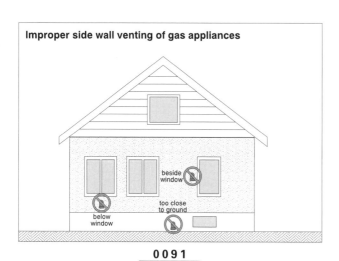

beside
window

too close
to ground

below
window

0091

High efficiency furnaces have long heat exchangers

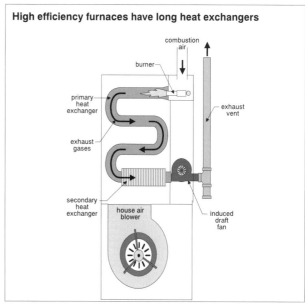

combustion
air

burner

primary
heat
exchanger

exhaust
gases

exhaust
vent

secondary
heat
exchanger

house air
blower

induced
draft
fan

0092

Condensation in high efficiency furnaces

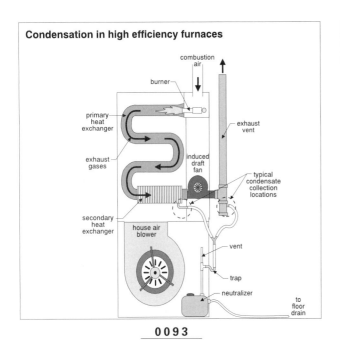

combustion air

burner

primary heat exchanger

exhaust gases

induced draft fan

exhaust vent

typical condensate collection locations

secondary heat exchanger

house air blower

vent

trap

neutralizer

to floor drain

0093

Maximizing temperature differences

in most high efficiency furnaces, heat is transferred from the coolest exhaust products to the coolest house air

this maximizes the temperature differential from beginning to end of the heat exchanger for the best flow of exhaust gases and also maximizes heat transfer to the house air

supply plenum

combustion air

hot house air

primary heat exchanger

exhaust gases

secondary heat exchanger

cool house air

house air blower

exhaust vent

induced draft fan

0094

Venting for high efficiency furnaces

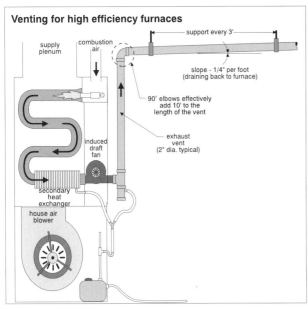

supply plenum

combustion air

support every 3'

slope - 1/4" per foot (draining back to furnace)

90° elbows effectively add 10' to the length of the vent

exhaust vent (2" dia. typical)

induced draft fan

secondary heat exchanger

house air blower

0095

Improper sidewall vent locations - high efficiency furnaces

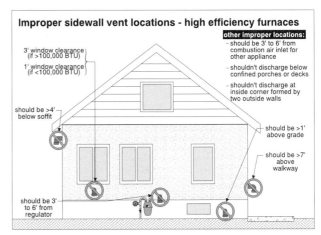

3' window clearance (if >100,000 BTU)

1' window clearance (if <100,000 BTU)

other improper locations:
– should be 3' to 6' from combustion air inlet for other appliance
– shouldn't discharge below confined porches or decks
– shouldn't discharge at inside corner formed by two outside walls

should be >4' below soffit

should be 3' to 6' from regulator

should be >1' above grade

should be >7' above walkway

0096

Safety devices for high efficiency furnaces

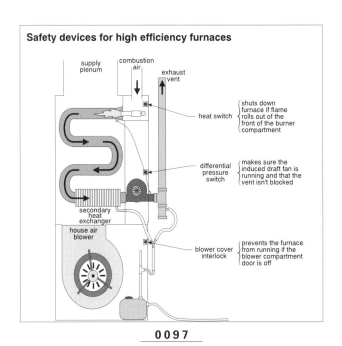

supply plenum · combustion air · exhaust vent

heat switch — shuts down furnace if flame rolls out of the front of the burner compartment

differential pressure switch — makes sure the induced draft fan is running and that the vent isn't blocked

secondary heat exchanger

house air blower

blower cover interlock — prevents the furnace from running if the blower compartment door is off

0097

Pulse furnace heat exchanger components

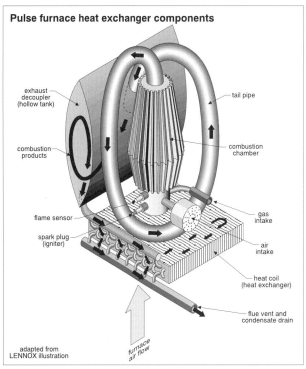

exhaust decoupler (hollow tank) · tail pipe · combustion products · combustion chamber · flame sensor · spark plug (igniter) · gas intake · air intake · heat coil (heat exchanger) · flue vent and condensate drain · furnace air flow

adapted from LENNOX illustration

0098

High efficiency furnaces and duct sizes

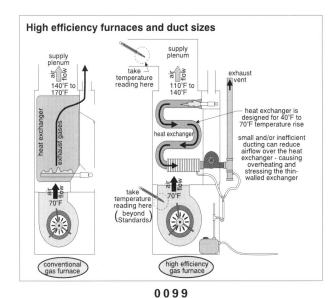

supply plenum · air flow 140°F to 170°F · take temperature reading here · supply plenum · air flow 110°F to 140°F · exhaust vent

heat exchanger · exhaust gases

heat exchanger is designed for 40°F to 70°F temperature rise

small and/or inefficient ducting can reduce airflow over the heat exchanger - causing overheating and stressing the thin-walled exchanger

air flow 70°F · take temperature reading here (beyond Standards)

conventional gas furnace · high efficiency gas furnace

0099

Automatic vent dampers

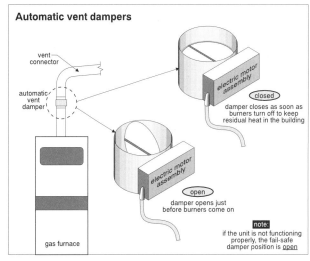

vent connector · automatic vent damper · electric motor assembly · closed — damper closes as soon as burners turn off to keep residual heat in the building

electric motor assembly · open — damper opens just before burners come on

gas furnace

note: if the unit is not functioning properly, the fail-safe damper position is open

0100

Inspecting high efficiency furnaces

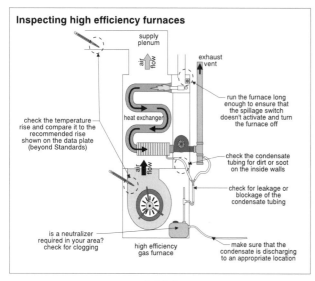

supply plenum

air flow

exhaust vent

heat exchanger

check the temperature rise and compare it to the recommended rise shown on the data plate (beyond Standards)

run the furnace long enough to ensure that the spillage switch doesn't activate and turn the furnace off

air flow

check the condensate tubing for dirt or soot on the inside walls

check for leakage or blockage of the condensate tubing

is a neutralizer required in your area? check for clogging

high efficiency gas furnace

make sure that the condensate is discharging to an appropriate location

0101

Combination furnace/water heater system

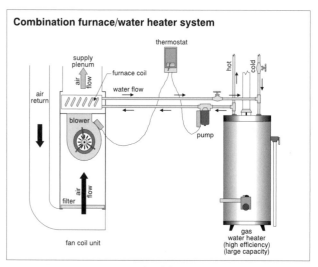

thermostat

supply plenum

furnace coil

air return

air flow

water flow

hot   cold

blower

pump

filter

air flow

fan coil unit

gas water heater (high efficiency) (large capacity)

0102

Tempering valve - combination system

in order to improve the efficiency of the fan coil unit, the water heater temperature is sometimes turned way up

a tempering valve adds a little cold water in with the hot (downstream of the fan coil) so that it is suitable for domestic use

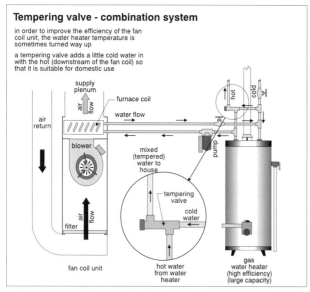

supply plenum

furnace coil

air return

air flow

water flow

hot   cold

blower

mixed (tempered) water to house

pump

tempering valve

cold water

filter

air flow

fan coil unit

hot water from water heater

gas water heater (high efficiency) (large capacity)

0103

Buried oil storage tank

clues that the oil tank is buried outside:

- you can't find one inside!

- pipe with oil filter and/or shut off valve emerging from foundation wall

- filler pipe and/or vent pipe is not right next to the wall

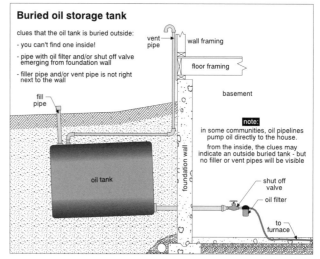

vent pipe

wall framing

floor framing

fill pipe

basement

note:

in some communities, oil pipelines pump oil directly to the house.

from the inside, the clues may indicate an outside buried tank - but no filler or vent pipes will be visible

oil tank

foundation wall

shut off valve

oil filter

to furnace

0104

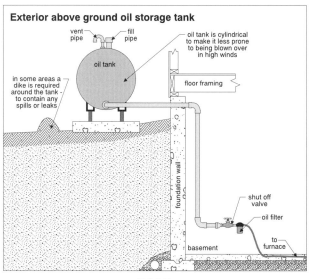

Exterior above ground oil storage tank

vent pipe

fill pipe

oil tank

oil tank is cylindrical to make it less prone to being blown over in high winds

floor framing

in some areas a dike is required around the tank - to contain any spills or leaks

foundation wall

shut off valve

oil filter

basement

to furnace

0105

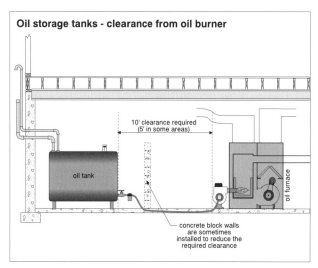

Oil storage tanks - clearance from oil burner

10' clearance required (5' in some areas)

oil tank

oil furnace

concrete block walls are sometimes installed to reduce the required clearance

0106

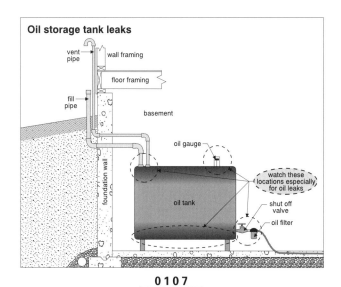

Oil storage tank leaks

vent pipe

wall framing

floor framing

fill pipe

basement

oil gauge

watch these locations especially for oil leaks

oil tank

shut off valve

oil filter

foundation wall

0107

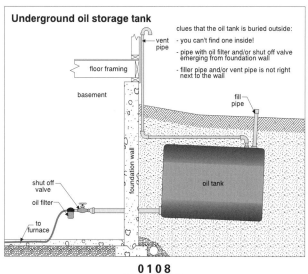

Underground oil storage tank

clues that the oil tank is buried outside:

- you can't find one inside!

- pipe with oil filter and/or shut off valve emerging from foundation wall

- filler pipe and/or vent pipe is not right next to the wall

vent pipe

floor framing

basement

fill pipe

shut off valve

oil filter

to furnace

foundation wall

oil tank

0108

Fill and vent piping

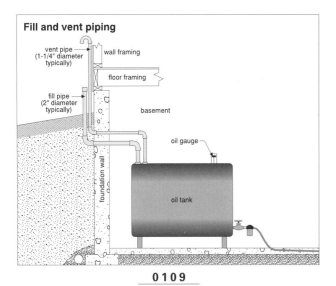

vent pipe
(1-1/4" diameter
typically)

wall framing

floor framing

fill pipe
(2" diameter
typically)

basement

oil gauge

foundation wall

oil tank

0 1 0 9

Caps for fill and vent pipes on oil storage tanks

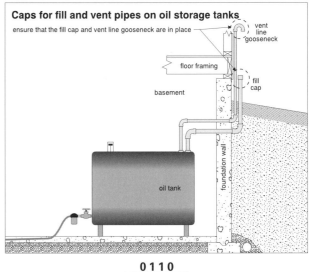

ensure that the fill cap and vent line gooseneck are in place

vent
line
gooseneck

floor framing

fill
cap

basement

foundation wall

oil tank

0 1 1 0

Undersized fill lines

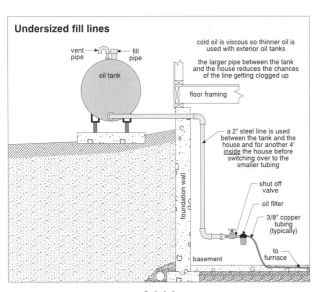

vent
pipe

fill
pipe

oil tank

cold oil is viscous so thinner oil is
used with exterior oil tanks

the larger pipe between the tank
and the house reduces the chances
of the line getting clogged up

floor framing

a 2" steel line is used
between the tank and the
house and for another 4'
<u>inside</u> the house before
switching over to the
smaller tubing

foundation wall

shut off
valve

oil filter

3/8" copper
tubing
(typically)

basement

to
furnace

0 1 1 1

Oil furnace emergency shutoffs

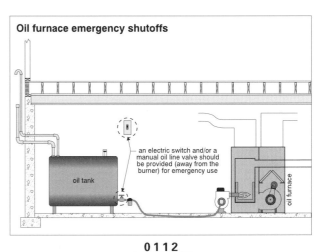

an electric switch and/or a
manual oil line valve should
be provided (away from the
burner) for emergency use

oil tank

oil furnace

0 1 1 2

Atomizing oil burner

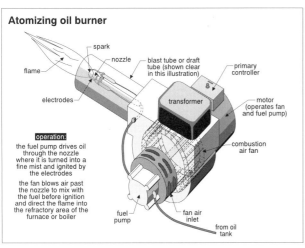

spark

nozzle

blast tube or draft
tube (shown clear
in this illustration)

primary
controller

flame

electrodes

transformer

motor
(operates fan
and fuel pump)

combustion
air fan

operation:
the fuel pump drives oil
through the nozzle
where it is turned into a
fine mist and ignited by
the electrodes

the fan blows air past
the nozzle to mix with
the fuel before ignition
and direct the flame into
the refractory area of the
furnace or boiler

fuel
pump

fan air
inlet

from oil
tank

0 1 1 3

Oil burner with flame retention head

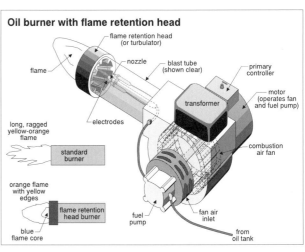

flame retention head
(or turbulator)

nozzle

blast tube
(shown clear)

primary
controller

flame

electrodes

transformer

motor
(operates fan
and fuel pump)

long, ragged
yellow-orange
flame

standard
burner

combustion
air fan

orange flame
with yellow
edges

flame retention
head burner

blue
flame core

fuel
pump

fan air
inlet

from
oil tank

0 1 1 4

Primary controller

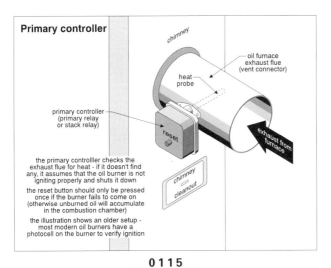

chimney

oil furnace exhaust flue (vent connector)

heat probe

primary controller (primary relay or stack relay)

reset

exhaust from furnace

chimney cleanout

the primary controller checks the exhaust flue for heat - if it doesn't find any, it assumes that the oil burner is not igniting properly and shuts it down

the reset button should only be pressed once if the burner fails to come on (otherwise unburned oil will accumulate in the combustion chamber)

the illustration shows an older setup - most modern oil burners have a photocell on the burner to verify ignition

0115

Inspecting an oil burner

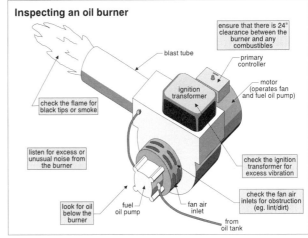

ensure that there is 24" clearance between the burner and any combustibles

blast tube

primary controller

ignition transformer

motor (operates fan and fuel oil pump)

check the flame for black tips or smoke

check the ignition transformer for excess vibration

listen for excess or unusual noise from the burner

check the fan air inlets for obstruction (eg. lint/dirt)

look for oil below the burner

fuel oil pump

fan air inlet

from oil tank

0116

Flame sensor

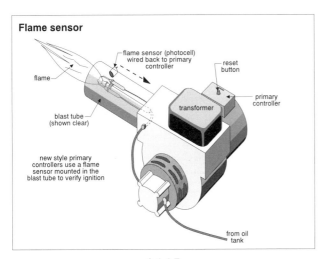

flame sensor (photocell) wired back to primary controller

reset button

flame

primary controller

transformer

blast tube (shown clear)

new style primary controllers use a flame sensor mounted in the blast tube to verify ignition

from oil tank

0117

Barometric damper
(draft regulator)

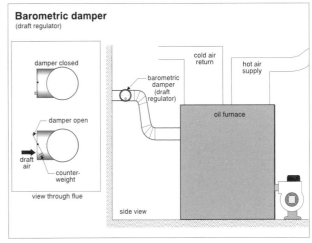

damper closed

damper open

draft air

counter-weight

view through flue

cold air return

hot air supply

barometric damper (draft regulator)

oil furnace

side view

0118

Barometric damper (draft regulator) problems

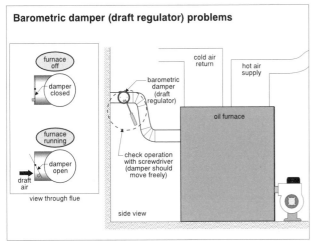

furnace off

damper closed

furnace running

damper open

draft air

view through flue

cold air return

hot air supply

barometric damper (draft regulator)

oil furnace

check operation with screwdriver (damper should move freely)

side view

0119

Spillage from barometric damper

damper stuck open

view through flue

exhaust gases leaking into house

cold air return

hot air supply

barometric damper (draft regulator)

oil furnace

check for spillage of exhaust gases here

side view

0120

Exhaust flue slope

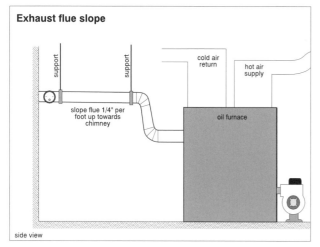

0121

Exhaust flue support

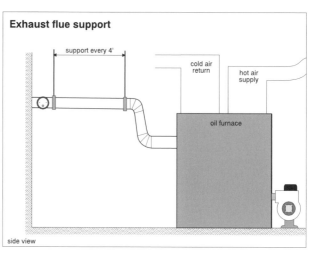

0122

Exhaust flue length

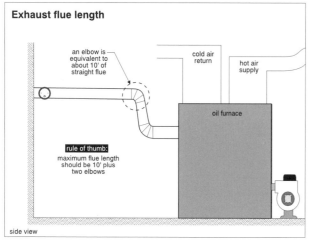

0123

Exhaust flue clearances

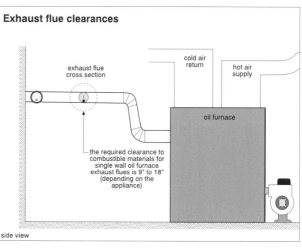

0124

Obstruction in exhaust flue

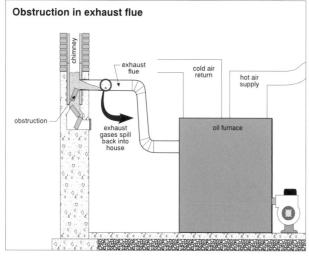

0125

Mid-efficiency oil furnace (sidewall vented)

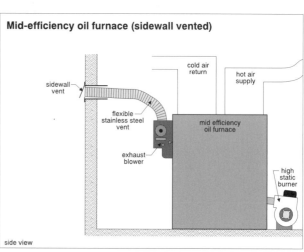

0126

Improper sidewall vent locations for oil furnaces

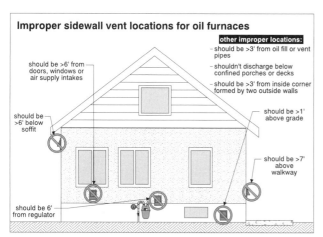

other improper locations:
- should be >3' from oil fill or vent pipes
- shouldn't discharge below confined porches or decks
- should be >3' from inside corner formed by two outside walls

should be >6' from doors, windows or air supply intakes

should be >6' below soffit

should be 6' from regulator

should be >1' above grade

should be >7' above walkway

0 1 2 7

How boilers work

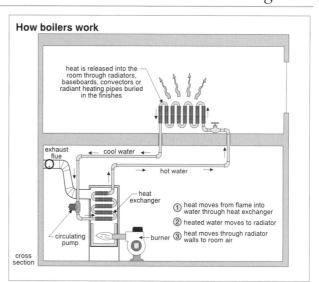

heat is released into the room through radiators, baseboards, convectors or radiant heating pipes buried in the finishes

exhaust flue

cool water

hot water

heat exchanger

circulating pump

burner

① heat moves from flame into water through heat exchanger
② heated water moves to radiator
③ heat moves through radiator walls to room air

cross section

0 1 2 8

How radiators heat the air through convection

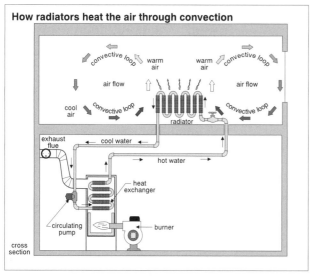

convective loop — warm air — warm air — convective loop

air flow — air flow

cool air — convective loop — convective loop

radiator

exhaust flue

cool water

hot water

heat exchanger

circulating pump

burner

cross section

0 1 2 9

Radiator covers

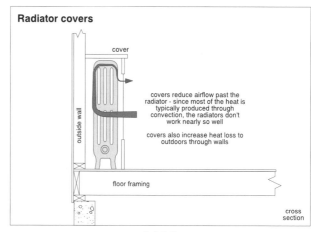

cover

outside wall

covers reduce airflow past the radiator - since most of the heat is typically produced through convection, the radiators don't work nearly so well

covers also increase heat loss to outdoors through walls

floor framing

cross section

0 1 3 0

Superheated water

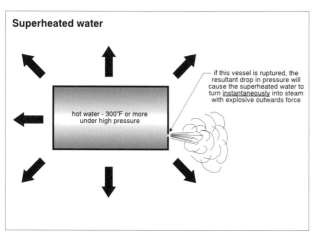

hot water - 300°F or more under high pressure

if this vessel is ruptured, the resultant drop in pressure will cause the superheated water to turn _instantaneously_ into steam with explosive outwards force

0 1 3 1

Heat exchangers

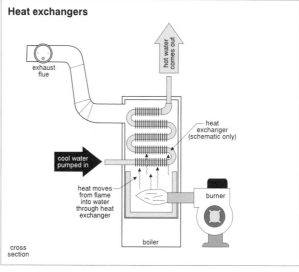

exhaust flue

hot water comes out

heat exchanger (schematic only)

cool water pumped in

heat moves from flame into water through heat exchanger

burner

boiler

cross section

0 1 3 2

Copper tube heat exchangers

these heat exchangers are prone to corrosion and build-up of deposits between the fins (fire side)

they also require good water flow through the exchanger to keep from overheating (leading to premature failure) - make sure the circulating pump is operating

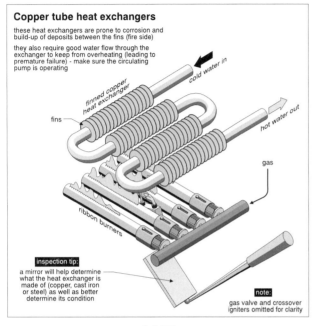

cold water in

finned copper heat exchanger

fins

hot water out

gas

ribbon burners

inspection tip:
a mirror will help determine what the heat exchanger is made of (copper, cast iron or steel) as well as better determine its condition

note:
gas valve and crossover igniters omitted for clarity

0 1 3 3

Pressure relief valve

note:
mechanism simplified for clarity

manual relief lever

spring

to drain   to drain

poppet valve

pressure less than valve rating

pressure greater than than valve rating (or manual override)

pressure relief valves come with specific pressure ratings (e.g. 30 psi for most boilers)

when the water pressure exceeds this level, the spring pressure is overcome and the valve is forced slightly open allowing excess pressure (and water) to escape

the valve can also be opened manually by flipping up the lever on the top but, this is not recommended on inspections

BTU rating of valve should be at least equal to boiler BTU rating

0 1 3 4

Pressure relief valve location

pressure relief valves are typically located at the top of boilers

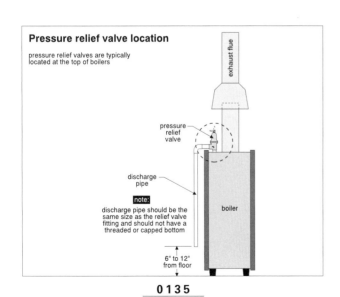

exhaust flue

pressure relief valve

discharge pipe

note:
discharge pipe should be the same size as the relief valve fitting and should not have a threaded or capped bottom

boiler

6" to 12" from floor

0 1 3 5

Inspecting pressure relief valves

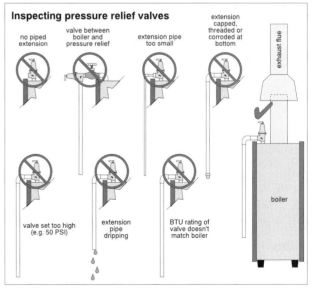

no piped extension

valve between boiler and pressure relief

extension pipe too small

extension capped, threaded or corroded at bottom

exhaust flue

boiler

valve set too high (e.g. 50 PSI)

extension pipe dripping

BTU rating of valve doesn't match boiler

0 1 3 6

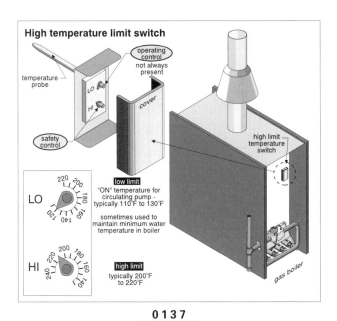

High temperature limit switch

temperature probe

operating control
not always present

LO

HI

cover

safety control

high limit temperature switch

low limit
"ON" temperature for circulating pump - typically 110°F to 130°F

sometimes used to maintain minimum water temperature in boiler

LO
220 200 180 160 140 120

HI
240 220 200 180 160 140

high limit
typically 200°F to 220°F

gas boiler

0137

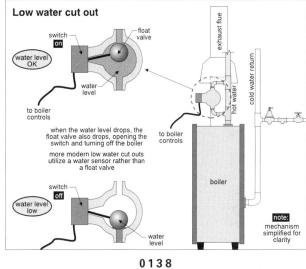

Low water cut out

switch
on

water level OK

float valve

water level

to boiler controls

when the water level drops, the float valve also drops, opening the switch and turning off the boiler

more modern low water cut outs utilize a water sensor rather than a float valve

switch
off

water level low

water level

to boiler controls

exhaust flue

hot water

cold water return

boiler

note:
mechanism simplified for clarity

0138

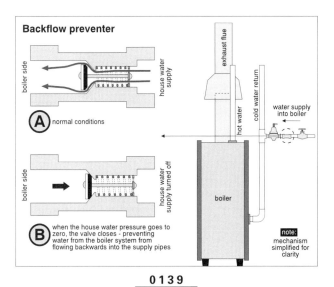

Backflow preventer

boiler side

house water supply

A normal conditions

boiler side

house water supply turned off

B when the house water pressure goes to zero, the valve closes - preventing water from the boiler system from flowing backwards into the supply pipes

exhaust flue

hot water

cold water return

water supply into boiler

boiler

note:
mechanism simplified for clarity

0139

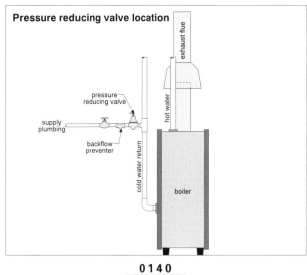

Pressure reducing valve location

exhaust flue

pressure reducing valve

supply plumbing

backflow preventer

hot water

cold water return

boiler

0140

Backflow preventer installed backwards

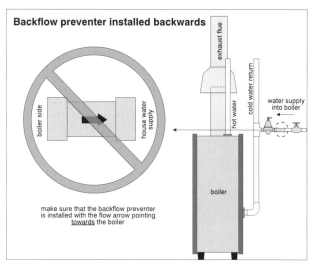

make sure that the backflow preventer
is installed with the flow arrow pointing
<u>towards</u> the boiler

0 1 4 1

Temperature and pressure gauge

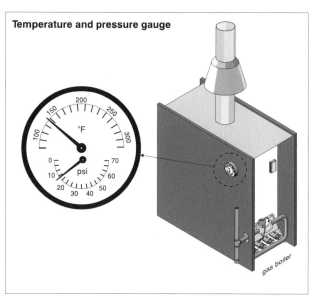

0 1 4 2

Pressure reducing valve

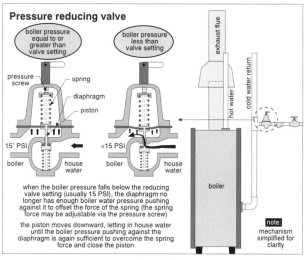

when the boiler pressure falls below the reducing
valve setting (usually 15 PSI), the diaphragm no
longer has enough boiler water pressure pushing
against it to offset the force of the spring (the spring
force may be adjustable via the pressure screw)

the piston moves downward, letting in house water
until the boiler pressure pushing against the
diaphragm is again sufficient to overcome the spring
force and close the piston

note:
mechanism
simplified for
clarity

0 1 4 3

Combined pressure reducing valve and pressure relief valve

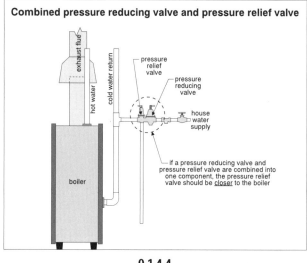

if a pressure reducing valve and
pressure relief valve are combined into
one component, the pressure relief
valve should be <u>closer</u> to the boiler

0 1 4 4

Pressure set too low

for a 3 story house, a pressure
reducing valve setting of 12 psi
won't be enough to push water
up into the 3rd floor radiators -
15 psi is just barely enough

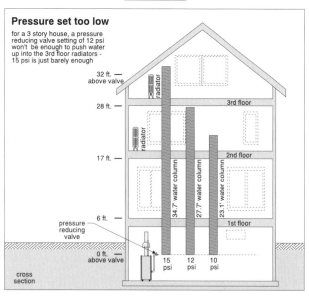

cross
section

0 1 4 5

Don't operate air bleed valves

0 1 4 6

Pressure reducing/relief valve installed backwards

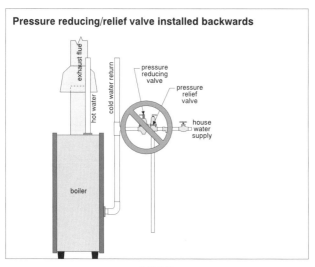

exhaust flue
hot water
cold water return
pressure reducing valve
pressure relief valve
house water supply
boiler

0 1 4 7

Air separators

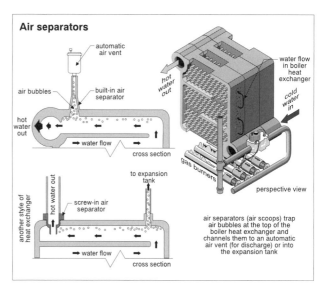

automatic air vent
air bubbles
built-in air separator
hot water out
water flow
cross section
hot water out
water flow in boiler heat exchanger
cold water in
gas burners
perspective view

another style of heat exchanger
hot water out
screw-in air separator
to expansion tank
water flow
cross section

air separators (air scoops) trap air bubbles at the top of the boiler heat exchanger and channels them to an automatic air vent (for discharge) or into the expansion tank

0 1 4 8

Aquastat - primary control

an aquastat is a primary control that is typically strapped to the hot water piping above the boiler

it must be tightly secured to the pipe to function properly

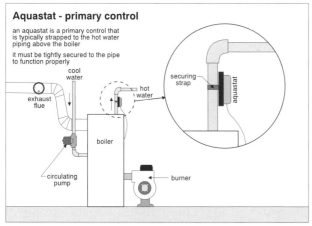

exhaust flue
cool water
hot water
securing strap
aquastat
circulating pump
boiler
burner

0 1 4 9

Inoperative aquastat

if the aquastat is loose or defective the boiler will not operate

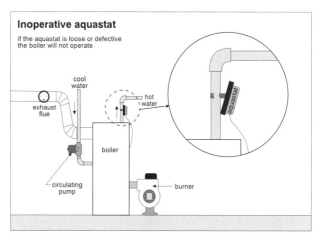

exhaust flue
cool water
hot water
aquastat
circulating pump
boiler
burner

0 1 5 0

Pump control

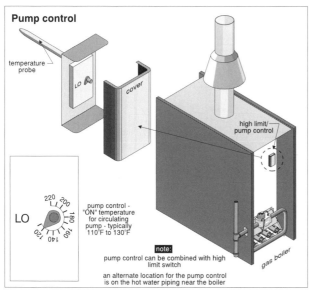

temperature probe
LO
cover
high limit/pump control
gas boiler

LO
220 200 180 160 140 120

pump control - "ON" temperature for circulating pump - typically 110°F to 130°F

note:
pump control can be combined with high limit switch

an alternate location for the pump control is on the hot water piping near the boiler

0 1 5 1

Zone control with pumps

zone 1
zone 2
thermostat
thermostat
hot water
cool water
boiler
circulating pumps
expansion tank not shown
cross section

0 1 5 2

Zone control with valves

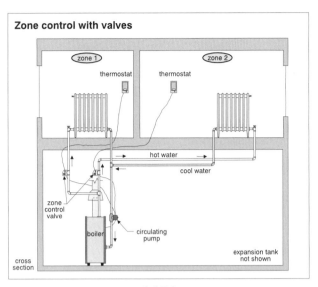

zone 1

zone 2

thermostat

thermostat

hot water

cool water

zone control valve

circulating pump

boiler

expansion tank not shown

cross section

0153

Outdoor air temperature sensor

exterior temperature probe and connecting wire

air temperature sensor

foundation wall

boiler

basement

temperature setting

cross section

0154

Flow control valves

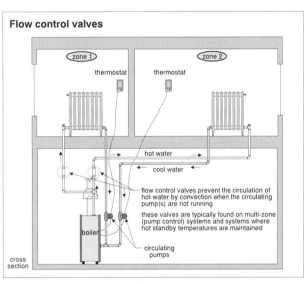

zone 1

zone 2

thermostat

thermostat

hot water

cool water

flow control valves prevent the circulation of hot water by convection when the circulating pump(s) are not running

these valves are typically found on multi-zone (pump control) systems and systems where hot standby temperatures are maintained

circulating pumps

boiler

cross section

0155

Open hydronic system

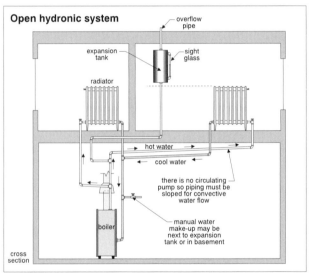

overflow pipe

expansion tank

sight glass

radiator

hot water

cool water

there is no circulating pump so piping must be sloped for convective water flow

manual water make-up may be next to expansion tank or in basement

boiler

cross section

0156

Closed hydronic system

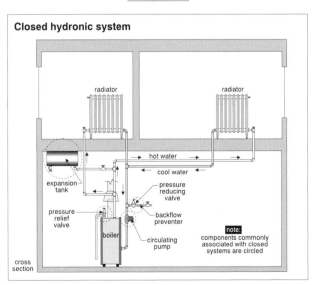

radiator

radiator

hot water

cool water

expansion tank

pressure reducing valve

backflow preventer

pressure relief valve

boiler

circulating pump

note:
components commonly associated with closed systems are circled

cross section

0157

Series loop

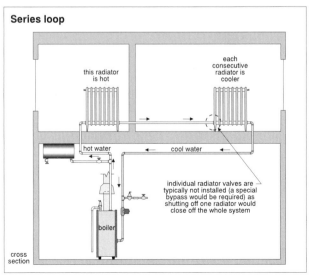

this radiator is hot

each consecutive radiator is cooler

hot water

cool water

individual radiator valves are typically not installed (a special bypass would be required) as shutting off one radiator would close off the whole system

boiler

cross section

0158

One-pipe system

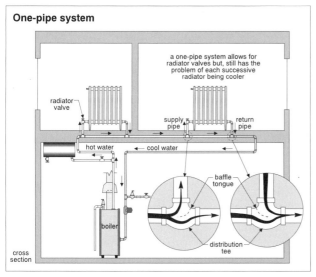

a one-pipe system allows for radiator valves but, still has the problem of each successive radiator being cooler

radiator valve

supply pipe

return pipe

hot water

cool water

baffle tongue

boiler

distribution tee

cross section

0159

Two-pipe system (direct return)

radiator

radiator

supply pipe

return pipe

expansion tank

hot water

cool water

boiler

cross section

0160

Two-pipe system (reverse return)

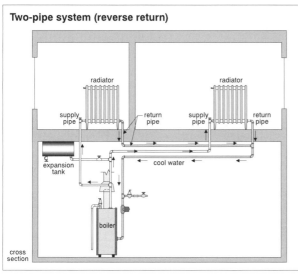

radiator

radiator

supply pipe

return pipe

supply pipe

return pipe

expansion tank

cool water

boiler

cross section

0161

Balancing methods

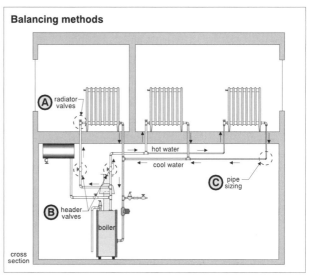

Ⓐ radiator valves

hot water

cool water

Ⓒ pipe sizing

Ⓑ header valves

boiler

cross section

0162

Expansion tank water levels

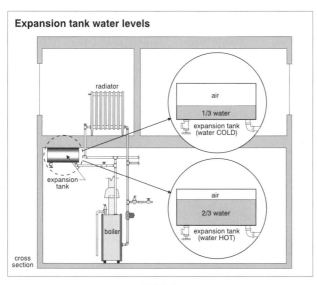

radiator

air

1/3 water

expansion tank (water COLD)

expansion tank

air

2/3 water

expansion tank (water HOT)

boiler

cross section

0163

Overflow pipe

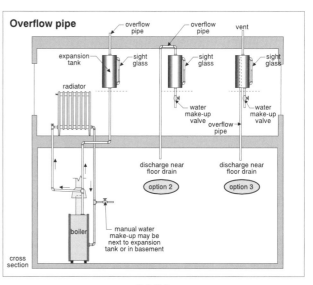

overflow pipe

overflow pipe

vent

expansion tank

sight glass

sight glass

sight glass

radiator

water make-up valve

water make-up valve

overflow pipe

discharge near floor drain

discharge near floor drain

option 2

option 3

boiler

manual water make-up may be next to expansion tank or in basement

cross section

0164

Conventional expansion tank

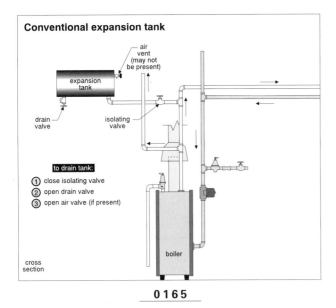

air vent (may not be present)

expansion tank

drain valve

isolating valve

to drain tank:
① close isolating valve
② open drain valve
③ open air valve (if present)

boiler

cross section

0165

Diaphragm tank

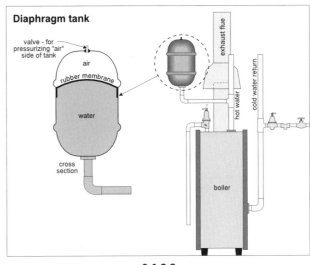

valve - for pressurizing "air" side of tank

exhaust flue

air

rubber membrane

water

hot water

cold water return

cross section

boiler

0166

Circulating pump

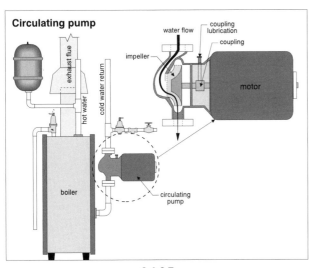

water flow

coupling lubrication

coupling

impeller

motor

exhaust flue

hot water

cold water return

boiler

circulating pump

0167

Pipe corrosion

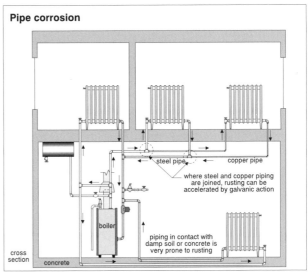

steel pipe

copper pipe

where steel and copper piping are joined, rusting can be accelerated by galvanic action

boiler

piping in contact with damp soil or concrete is very prone to rusting

cross section

concrete

0168

Extending hot water systems

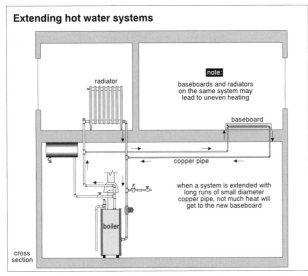

radiator

note:
baseboards and radiators on the same system may lead to uneven heating

baseboard

copper pipe

when a system is extended with long runs of small diameter copper pipe, not much heat will get to the new baseboard

boiler

cross section

0169

Covering radiators reduces efficiency

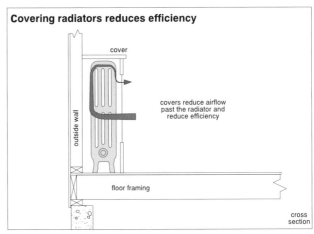

cover

outside wall

covers reduce airflow past the radiator and reduce efficiency

floor framing

cross section

0170

Convector

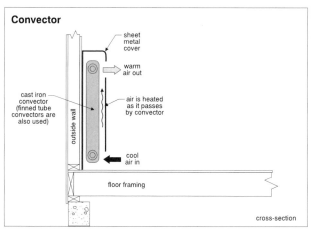

- sheet metal cover
- warm air out
- cast iron convector (finned tube convectors are also used)
- air is heated as it passes by convector
- outside wall
- cool air in
- floor framing

cross-section

0 1 7 1

Finned tube baseboard

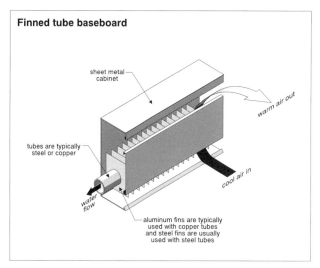

- sheet metal cabinet
- warm air out
- tubes are typically steel or copper
- cool air in
- water flow
- aluminum fins are typically used with copper tubes and steel fins are usually used with steel tubes

0 1 7 2

Cast iron baseboard

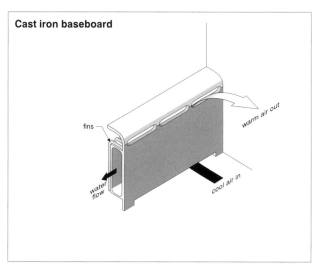

- warm air out
- fins
- cool air in
- water flow

0 1 7 3

Mixing systems

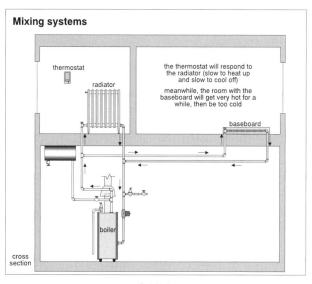

- thermostat
- radiator
- the thermostat will respond to the radiator (slow to heat up and slow to cool off)
- meanwhile, the room with the baseboard will get very hot for a while, then be too cold
- baseboard
- boiler

cross section

0 1 7 4

Radiators on ceilings or high on walls

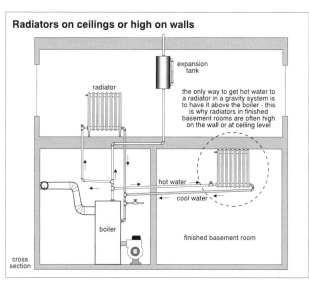

- expansion tank
- radiator
- the only way to get hot water to a radiator in a gravity system is to have it above the boiler - this is why radiators in finished basement rooms are often high on the wall or at ceiling level
- hot water
- cool water
- boiler
- finished basement room

cross section

0 1 7 5

Radiator valve leaks

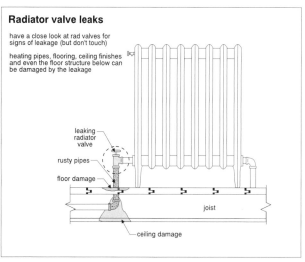

have a close look at rad valves for signs of leakage (but don't touch)

heating pipes, flooring, ceiling finishes and even the floor structure below can be damaged by the leakage

- leaking radiator valve
- rusty pipes
- floor damage
- joist
- ceiling damage

0 1 7 6

Hot water radiant heat

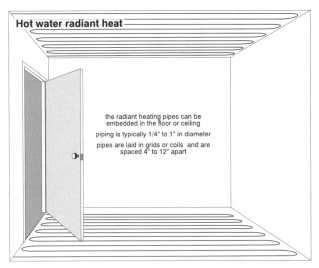

the radiant heating pipes can be embedded in the floor or ceiling

piping is typically 1/4" to 1" in diameter

pipes are laid in grids or coils and are spaced 4" to 12" apart

0177

Water blender on radiant system

since radiant heating systems need to use cooler water than conventional systems, a water blender (tempering valve) is often used to mix some of the cool returning water with the hot water coming off the boiler

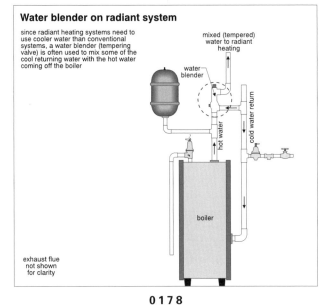

mixed (tempered) water to radiant heating

water blender

hot water

cold water return

boiler

exhaust flue not shown for clarity

0178

Tankless coil

a tankless coil uses the hot boiler water to heat water for the supply plumbing

it is a slide-in option for some boilers

to transfer heat from the <u>hottest</u> water, it is located near the top of the boiler

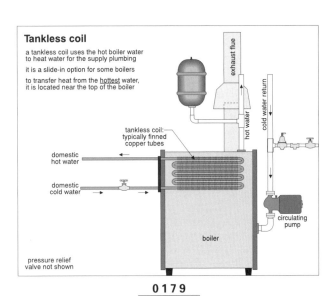

exhaust flue

hot water

cold water return

tankless coil: typically finned copper tubes

domestic hot water

domestic cold water

circulating pump

boiler

pressure relief valve not shown

0179

Side arm heater

like a tankless coil, a side arm heater uses the hot boiler water to heat water for the supply plumbing

side arm heaters, however, are mounted outside the boiler

hot water from the boiler is drawn across the side arm coil and then returned to the boiler

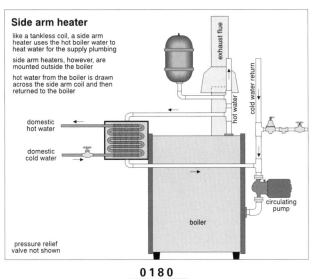

exhaust flue

hot water

cold water return

domestic hot water

domestic cold water

circulating pump

boiler

pressure relief valve not shown

0180

Spillage switch

if a spillage switch senses high temperatures at the draft hood it will shut down the burner

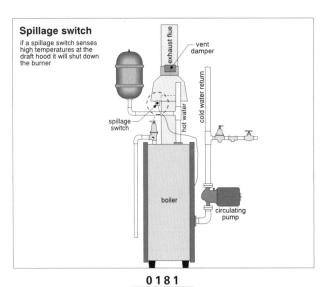

exhaust flue

vent damper

hot water

cold water return

spillage switch

boiler

circulating pump

0181

Bypass loop

a bypass loop uses a mixing valve to direct some of the hot water coming off the boiler back into the cold water return to minimize thermal shock and condensation on the combustion side of the heat exchanger

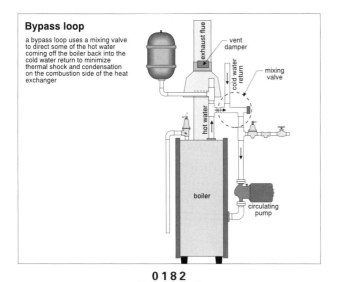

0182

Pulse combustion - how it works

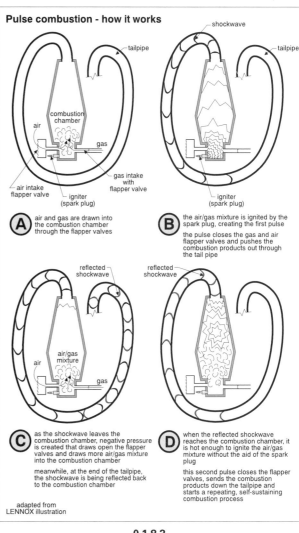

A air and gas are drawn into the combustion chamber through the flapper valves

B the air/gas mixture is ignited by the spark plug, creating the first pulse

the pulse closes the gas and air flapper valves and pushes the combustion products out through the tail pipe

C as the shockwave leaves the combustion chamber, negative pressure is created that draws open the flapper valves and draws more air/gas mixture into the combustion chamber

meanwhile, at the end of the tailpipe, the shockwave is being reflected back to the combustion chamber

D when the reflected shockwave reaches the combustion chamber, it is hot enough to ignite the air/gas mixture without the aid of the spark plug

this second pulse closes the flapper valves, sends the combustion products down the tailpipe and starts a repeating, self-sustaining combustion process

adapted from
LENNOX illustration

0183

Chimneys

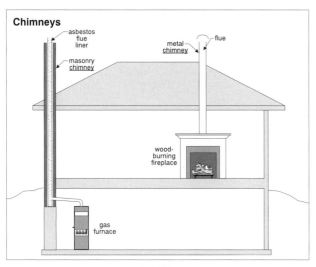

0184

Vents

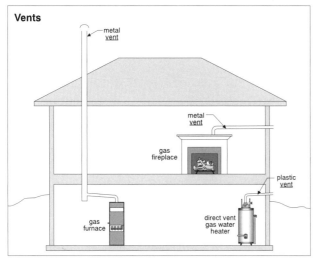

0185

Chimneys are not supporting structures

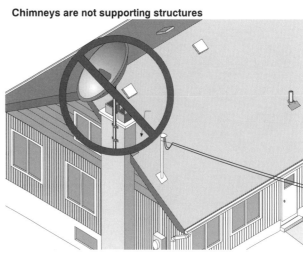

0186

Warm chimneys are best

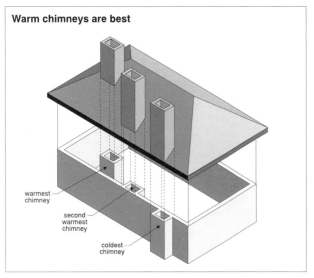

warmest chimney

second warmest chimney

coldest chimney

0187

Chimney extender

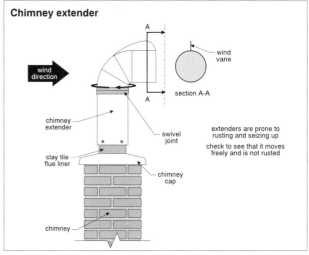

wind direction

A

A

section A-A

wind vane

chimney extender

swivel joint

clay tile flue liner

chimney cap

chimney

extenders are prone to rusting and seizing up

check to see that it moves freely and is not rusted

0188

Masonry chimneys

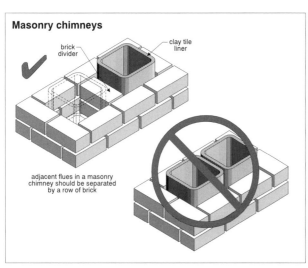

brick divider

clay tile liner

adjacent flues in a masonry chimney should be separated by a row of brick

0189

Basic masonry chimney components

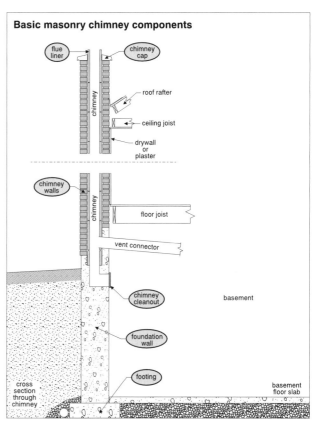

flue liner

chimney cap

chimney

roof rafter

ceiling joist

drywall or plaster

chimney walls

chimney

floor joist

vent connector

chimney cleanout

basement

foundation wall

footing

cross section through chimney

basement floor slab

0190

Basic masonry chimney and fireplace components

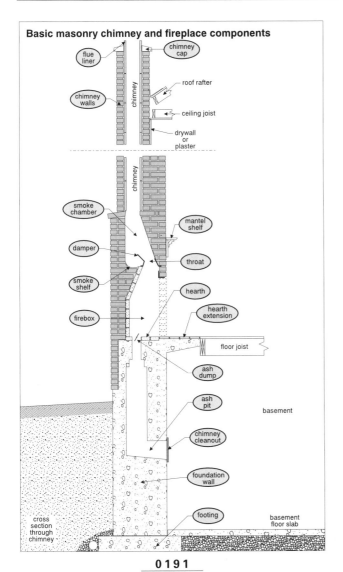

flue liner

chimney cap

chimney walls

roof rafter

ceiling joist

drywall or plaster

chimney

smoke chamber

mantel shelf

damper

throat

smoke shelf

hearth

firebox

hearth extension

floor joist

ash dump

ash pit

basement

chimney cleanout

foundation wall

footing

basement floor slab

cross section through chimney

0191

Chimney walls

chimney walls (and flue dividers) should be at least 3" to 4" (one brick) thick

the width of most flues is equal to the length of one brick - the total number of chimney flues can often be determined by counting bricks

brick divider

clay tile liner

3" to 4"

flue width

3" to 4"

flue width

3" to 4"

3" to 4"

0192

Lateral support for masonry chimneys

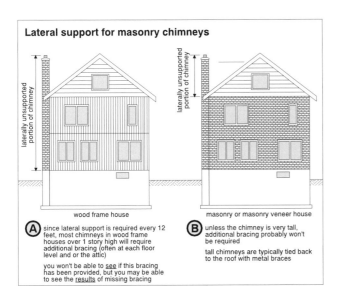

laterally unsupported portion of chimney

laterally unsupported portion of chimney

wood frame house

masonry or masonry veneer house

A since lateral support is required every 12 feet, most chimneys in wood frame houses over 1 story high will require additional bracing (often at each floor level and or the attic)

you won't be able to <u>see</u> if this bracing has been provided, but you may be able to see the <u>results</u> of missing bracing

B unless the chimney is very tall, additional bracing probably won't be required

tall chimneys are typically tied back to the roof with metal braces

0193

Clay tile flue liners

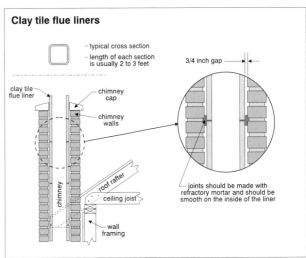

– typical cross section

– length of each section is usually 2 to 3 feet

3/4 inch gap

clay tile flue liner

chimney cap

chimney walls

chimney

roof rafter

ceiling joist

wall framing

joints should be made with refractory mortar and should be smooth on the inside of the liner

0194

Chimney offsets

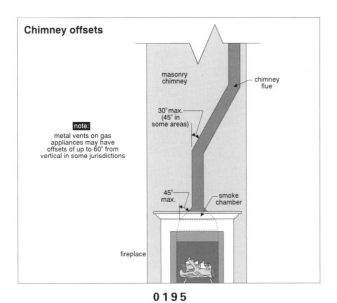

masonry chimney

chimney flue

30° max. (45° in some areas)

note: metal vents on gas appliances may have offsets of up to 60° from vertical in some jurisdictions

45° max.

smoke chamber

fireplace

0195

Miter the flue liners

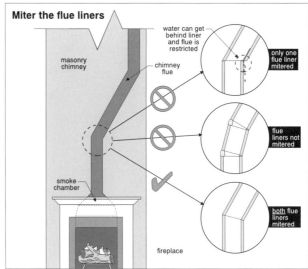

masonry chimney

chimney flue

water can get behind liner and flue is restricted

only one flue liner mitered

flue liners not mitered

smoke chamber

both flue liners mitered

fireplace

0196

Chimney clearances

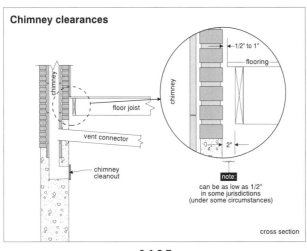

chimney

1/2" to 1"

flooring

chimney

floor joist

vent connector

chimney cleanout

2"

note: can be as low as 1/2" in some jurisdictions (under some circumstances)

cross section

0197

Chimneys supporting framing members

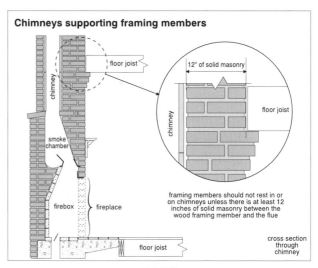

chimney

floor joist

12" of solid masonry

smoke chamber

chimney

floor joist

firebox

fireplace

framing members should not rest in or on chimneys unless there is at least 12 inches of solid masonry between the wood framing member and the flue

floor joist

cross section through chimney

0198

What makes a good chimney cap?

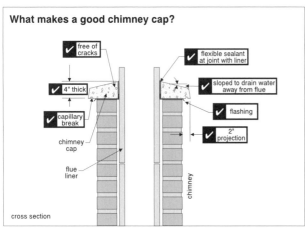

✔ free of cracks

✔ flexible sealant at joint with liner

✔ 4" thick

✔ sloped to drain water away from flue

✔ capillary break

✔ flashing

chimney cap

✔ 2" projection

flue liner

chimney

cross section

0199

Chimney deterioration

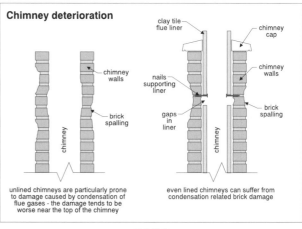

clay tile flue liner

chimney cap

chimney walls

nails supporting liner

chimney walls

brick spalling

gaps in liner

brick spalling

chimney

chimney

unlined chimneys are particularly prone to damage caused by condensation of flue gases - the damage tends to be worse near the top of the chimney

even lined chimneys can suffer from condensation related brick damage

0200

Removing abandoned chimneys

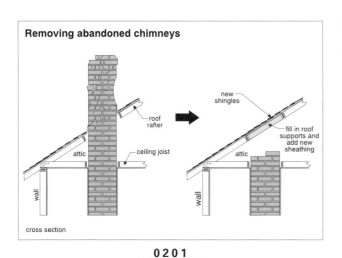

new shingles

roof rafter

fill in roof supports and add new sheathing

attic

ceiling joist

attic

wall

wall

cross section

__0201__

Causes of chimney settling or leaning

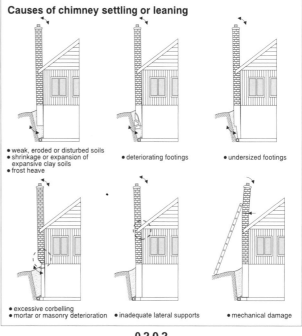

- weak, eroded or disturbed soils
- shrinkage or expansion of expansive clay soils
- frost heave

- deteriorating footings

- undersized footings

- excessive corbelling
- mortar or masonry deterioration

- inadequate lateral supports

- mechanical damage

__0202__

Tall masonry chimneys

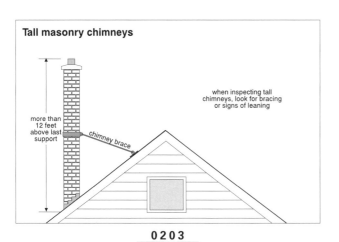

more than 12 feet above last support

chimney brace

when inspecting tall chimneys, look for bracing or signs of leaning

__0203__

3-sided chimneys

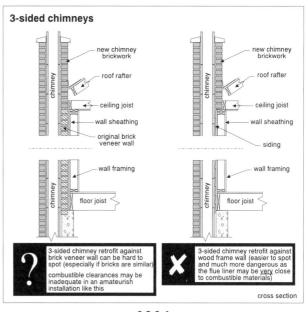

new chimney brickwork

chimney

roof rafter

ceiling joist

wall sheathing

original brick veneer wall

new chimney brickwork

chimney

roof rafter

ceiling joist

wall sheathing

siding

chimney

wall framing

floor joist

chimney

wall framing

floor joist

? 3-sided chimney retrofit against brick veneer wall can be hard to spot (especially if bricks are similar)

combustible clearances may be inadequate in an amateurish installation like this

✗ 3-sided chimney retrofit against wood frame wall (easier to spot and much more dangerous as the flue liner may be <u>very</u> close to combustible materials)

cross section

__0204__

Bracket chimneys

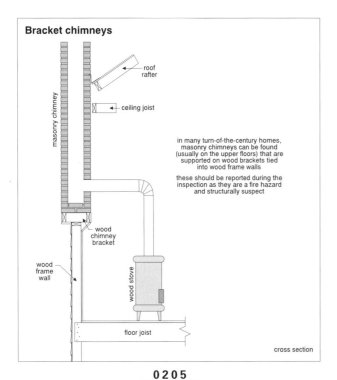

masonry chimney

roof rafter

ceiling joist

in many turn-of-the-century homes, masonry chimneys can be found (usually on the upper floors) that are supported on wood brackets tied into wood frame walls

these should be reported during the inspection as they are a fire hazard and structurally suspect

wood chimney bracket

wood frame wall

wood stove

floor joist

cross section

0205

Flue divider missing

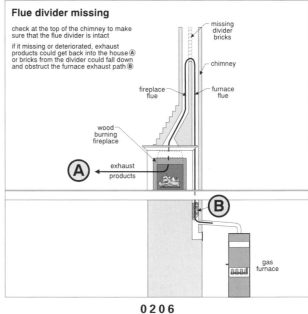

check at the top of the chimney to make sure that the flue divider is intact

if it missing or deteriorated, exhaust products could get back into the house Ⓐ or bricks from the divider could fall down and obstruct the furnace exhaust path Ⓑ

missing divider bricks

chimney

fireplace flue

furnace flue

wood burning fireplace

Ⓐ   exhaust products

Ⓑ

gas furnace

0206

Chimney leaning above roof line

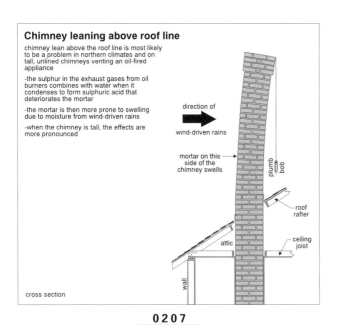

chimney lean above the roof line is most likely to be a problem in northern climates and on tall, unlined chimneys venting an oil-fired appliance

-the sulphur in the exhaust gases from oil burners combines with water when it condenses to form sulphuric acid that deteriorates the mortar

-the mortar is then more prone to swelling due to moisture from wind-driven rains

-when the chimney is tall, the effects are more pronounced

direction of

wind-driven rains

mortar on this side of the chimney swells

plumb bob

roof rafter

ceiling joist

attic

wall

cross section

0207

Proper chimney height

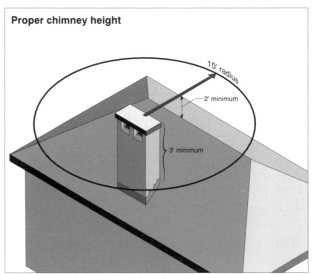

10' radius

2' minimum

3' minimum

0208

Chimney height above appliance

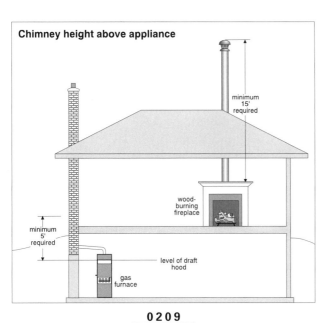

minimum 15' required

wood-burning fireplace

minimum 5' required

level of draft hood

gas furnace

0209

Fireplace cleanout - door too close to combustibles

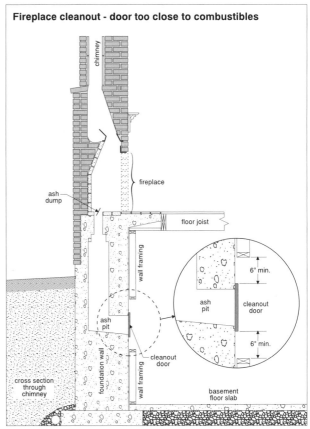

chimney

fireplace

ash dump

floor joist

wall framing

6" min.

ash pit

cleanout door

6" min.

ash pit

cleanout door

cross section through chimney

foundation wall

wall framing

basement floor slab

0210

Incomplete liner

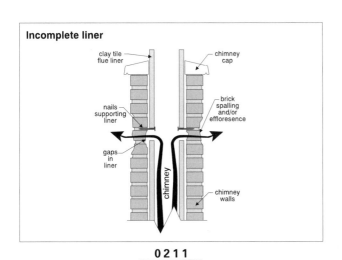

clay tile flue liner

chimney cap

nails supporting liner

brick spalling and/or effloresence

gaps in liner

chimney

chimney walls

0211

Fire stopping

you should not be able to see from one floor to the next around a masonry chimney

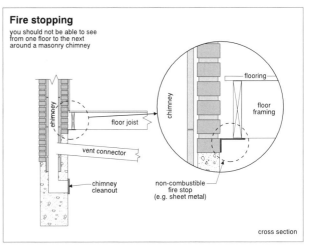

chimney

flooring

chimney

floor framing

floor joist

vent connector

non-combustible fire stop (e.g. sheet metal)

chimney cleanout

cross section

0212

Rain caps and spark arresters

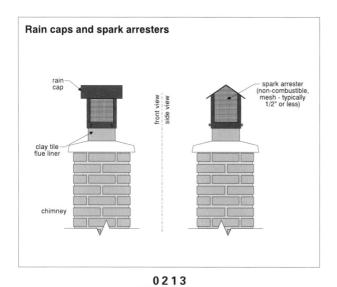

0 2 1 3

Improper slope on cap

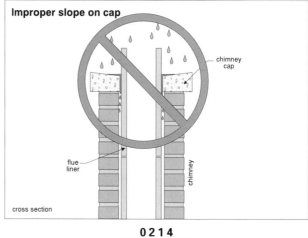

0 2 1 4

Drip edge on cap

cross section

0 2 1 5

Creosote deposits

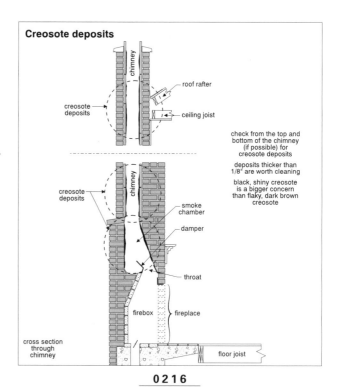

check from the top and bottom of the chimney (if possible) for creosote deposits

deposits thicker than 1/8" are worth cleaning

black, shiny creosote is a bigger concern than flaky, dark brown creosote

0 2 1 6

Vent connector loose at chimney

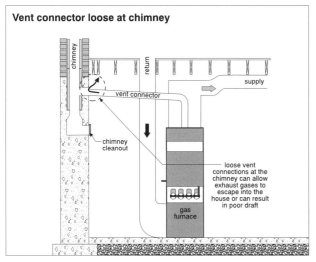

chimney

return

supply

vent connector

chimney cleanout

loose vent connections at the chimney can allow exhaust gases to escape into the house or can result in poor draft

gas furnace

0 2 1 7

Flue or vent connector obstructed

chimney

return

supply

vent connector

obstruction

12" min.

chimney cleanout

exhaust gases spill back into house

gas furnace

0 2 1 8

Abandoned flue openings

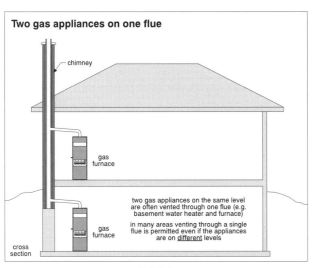

sheet metal plate

abandoned flue openings should be properly filled to prevent overheating or leakage of exhaust products

0 2 1 9

Two fireplaces on one flue

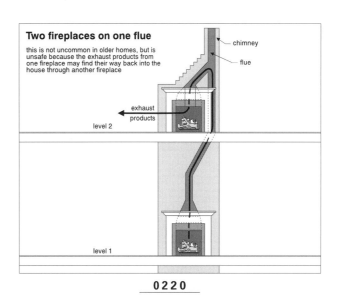

this is not uncommon in older homes, but is unsafe because the exhaust products from one fireplace may find their way back into the house through another fireplace

chimney

flue

exhaust products

level 2

level 1

0 2 2 0

Two gas appliances on one flue

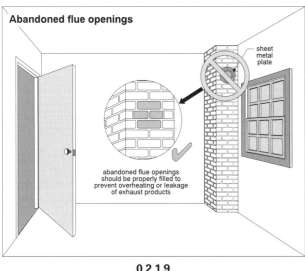

chimney

gas furnace

gas furnace

two gas appliances on the same level are often vented through one flue (e.g. basement water heater and furnace)

in many areas venting through a single flue is permitted even if the appliances are on <u>different</u> levels

cross section

0 2 2 1

Wood below oil on same floor level

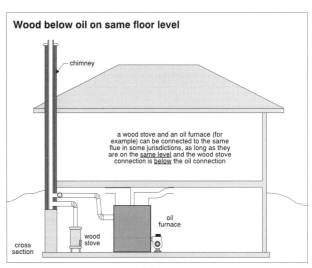

chimney

a wood stove and an oil furnace (for example) can be connected to the same flue in some jurisdictions, as long as they are on the <u>same level</u> and the wood stove connection is <u>below</u> the oil connection

oil furnace

wood stove

cross section

0 2 2 2

Type "B" vent

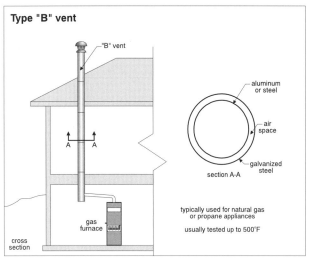

"B" vent

aluminum or steel

air space

galvanized steel

section A-A

A   A

gas furnace

typically used for natural gas or propane appliances

usually tested up to 500°F

cross section

0223

Type "L" vent

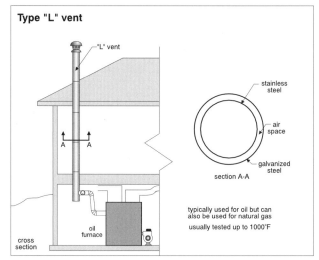

"L" vent

stainless steel

air space

galvanized steel

section A-A

A   A

oil furnace

typically used for oil but can also be used for natural gas

usually tested up to 1000°F

cross section

0224

Class "A" chimney

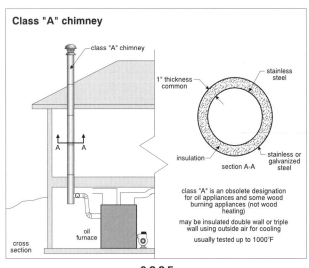

class "A" chimney

1" thickness common

stainless steel

insulation

stainless or galvanized steel

section A-A

A   A

oil furnace

class "A" is an obsolete designation for oil appliances and some wood burning appliances (not wood heating)

may be insulated double wall or triple wall using outside air for cooling

usually tested up to 1000°F

cross section

0225

Triple-wall metal chimney

triple-wall chimneys use air rather than insulation to cool the outer walls of the chimney

this type of chimney is suitable only for zero clearance fireplaces - the flue would get too cool if used with a wood burning stove, causing creosote build-up

some triple-wall chimneys do not circulate air through the two outer chambers - insulation is provided by the "dead air" space

the inner lining is typically stainless steel while the outer walls can be stainless steel, aluminum or galvanized steel

exhaust gas

warm air out

cold air in

exhaust gas

top of zero clearance fireplace

inner lining

0226

650°C chimneys

also called Super Chimneys or 629 Chimneys

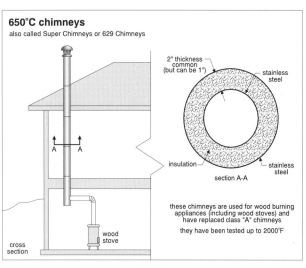

2" thickness common (but can be 1")

stainless steel

insulation

stainless steel

section A-A

A   A

wood stove

these chimneys are used for wood burning appliances (including wood stoves) and have replaced class "A" chimneys

they have been tested up to 2000°F

cross section

0227

Connections of metal vent pieces

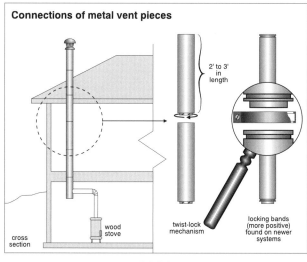

2' to 3' in length

wood stove

twist-lock mechanism

locking bands (more positive) found on newer systems

cross section

0228

Metal chimney bracing

chimneys over 5' to 6' tall
should be braced

check the brace for loose
or corroded supports

>5'-6'

chimney brace

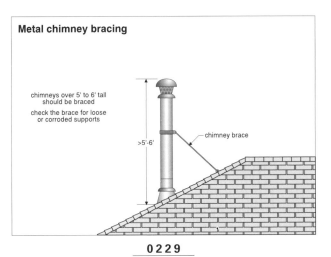

0 2 2 9

Metal chimney combustible clearances

1"

2" to 2-1/2"

2"

"B" vent and "L" vent

class "A" chimney

650 C chimney

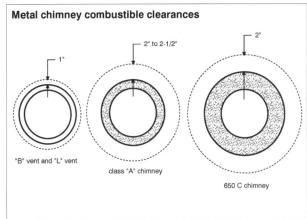

0 2 3 0

Fire stopping

collar can be used
with firestop to
provide support at
floor/ceiling level

metal
firestop
spacer

metal chimney

floor joist

2"

firestop
spacer shown
upside down

2"

cross
section

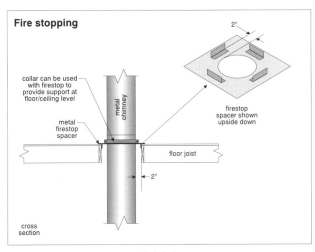

0 2 3 1

Missing chimney cap

a proper cap is particularly important
with a metal chimney/factory built
fireplace combination because there
is usually no smoke shelf to catch
water running down the chimney and
the metal flue is readily corroded by
water

chimney caps also provide protection
against downdrafts

rain   downdrafts

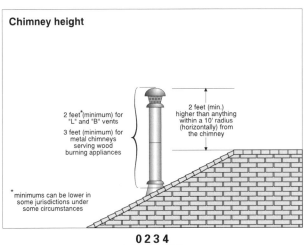

0 2 3 2

Rusting and/or pitting metal chimneys

pitting

rusting

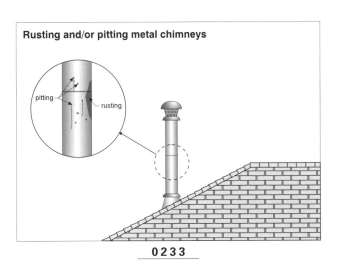

0 2 3 3

Chimney height

2 feet (minimum) for
"L" and "B" vents

3 feet (minimum) for
metal chimneys
serving wood
burning appliances

2 feet (min.)
higher than anything
within a 10' radius
(horizontally) from
the chimney

* minimums can be lower in
some jurisdictions under
some circumstances

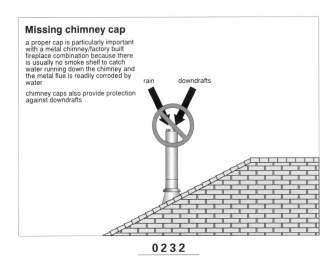

0 2 3 4

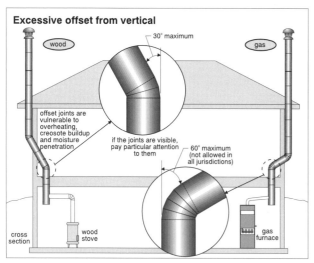

Excessive offset from vertical

wood

gas

30° maximum

offset joints are vulnerable to overheating, creosote buildup and moisture penetration

if the joints are visible, pay particular attention to them

60° maximum (not allowed in all jurisdictions)

cross section

wood stove

gas furnace

0235

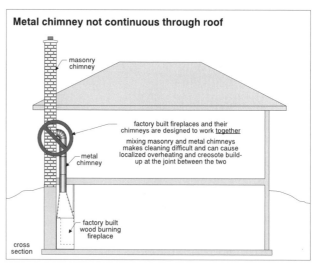

Metal chimney not continuous through roof

masonry chimney

factory built fireplaces and their chimneys are designed to work together

mixing masonry and metal chimneys makes cleaning difficult and can cause localized overheating and creosote build-up at the joint between the two

metal chimney

factory built wood burning fireplace

cross section

0236

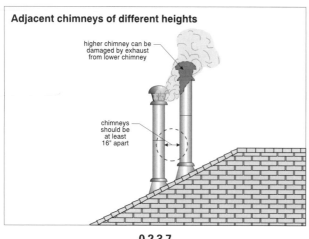

Adjacent chimneys of different heights

higher chimney can be damaged by exhaust from lower chimney

chimneys should be at least 16" apart

0237

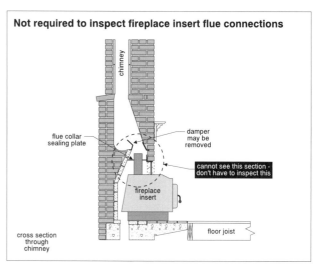

Not required to inspect fireplace insert flue connections

chimney

flue collar sealing plate

damper may be removed

cannot see this section - don't have to inspect this

fireplace insert

floor joist

cross section through chimney

0238

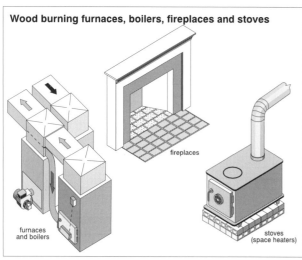

Wood burning furnaces, boilers, fireplaces and stoves

fireplaces

furnaces and boilers

stoves (space heaters)

0239

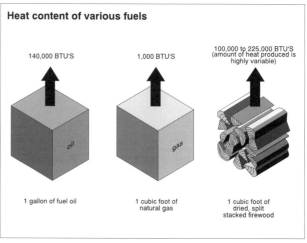

Heat content of various fuels

140,000 BTU'S

1,000 BTU'S

100,000 to 225,000 BTU'S (amount of heat produced is highly variable)

oil

gas

1 gallon of fuel oil

1 cubic foot of natural gas

1 cubic foot of dried, split stacked firewood

0240

Components of wood furnaces

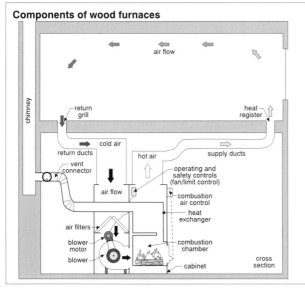

air flow

chimney

return grill

heat register

cold air

return ducts

hot air

supply ducts

vent connector

operating and safety controls (fan/limit control)

air flow

combustion air control

heat exchanger

air filters

blower motor

blower

combustion chamber

cabinet

cross section

0 2 4 1

Ventilation/cooling air for wood furnaces

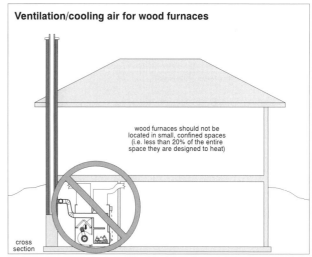

wood furnaces should not be located in small, confined spaces (i.e. less than 20% of the entire space they are designed to heat)

cross section

0 2 4 2

Combustion air damper arrangement

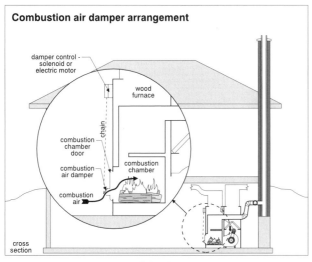

damper control - solenoid or electric motor

wood furnace

chain

combustion chamber door

combustion air damper

combustion chamber

combustion air

cross section

0 2 4 3

How a combustion air damper works

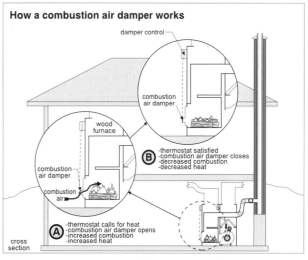

damper control

combustion air damper

wood furnace

combustion air damper

combustion air

(B) -thermostat satisfied
-combustion air damper closes
-decreased combustion
-decreased heat

(A) -thermostat calls for heat
-combustion air damper opens
-increased combustion
-increased heat

cross section

0 2 4 4

Forced draft combustion

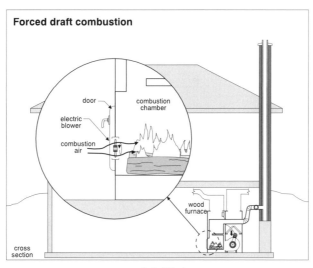

door

combustion chamber

electric blower

combustion air

wood furnace

cross section

0 2 4 5

Combustion chamber (firebox)

cold air

hot air

air flow

operating and safety controls

air filters

blower motor

blower

combustion chamber

walls of combustion chamber are typically heavy steel plate (1/4" thick)

on older units, the firebox may be lined with brick

wood furnace

cross section

can be designed to accommodate 16", 24" or 48" pieces of wood

0246

Two combustion chambers

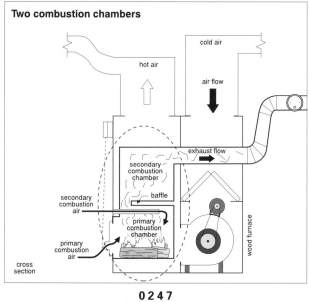

hot air

cold air

air flow

exhaust flow

secondary combustion chamber

secondary combustion air

baffle

primary combustion chamber

primary combustion air

wood furnace

cross section

0247

Heat exchangers shouldn't have nooks and crannys

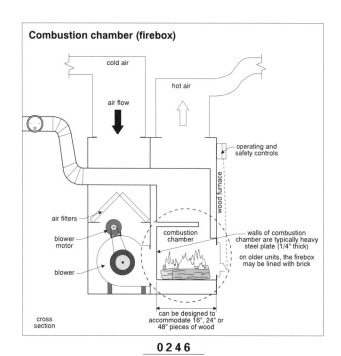

hot air

cold air

air flow

heat exchangers shouldn't have sharp elbows or nooks and crannys that can collect debris and be hard to clean

heat exchanger

exhaust flow

wood furnace

cross section

0248

Checking the barometric damper

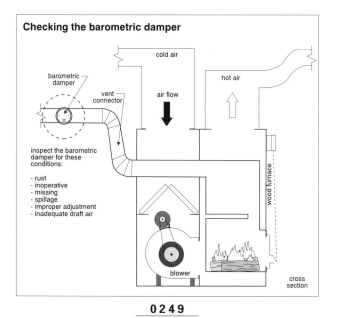

cold air

hot air

barometric damper

vent connector

air flow

inspect the barometric damper for these conditions:

- rust
- inoperative
- missing
- spillage
- improper adjustment
- inadequate draft air

wood furnace

blower

cross section

0249

Vent connector combustible clearances

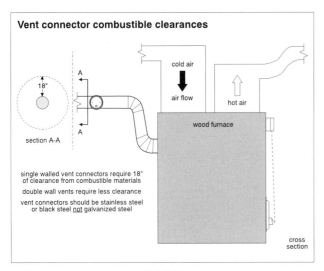

18"

A

A

section A-A

cold air
air flow

hot air

wood furnace

cross section

single walled vent connectors require 18"
of clearance from combustible materials

double wall vents require less clearance

vent connectors should be stainless steel
or black steel <u>not</u> galvanized steel

0 2 5 0

Joining vent connector sections

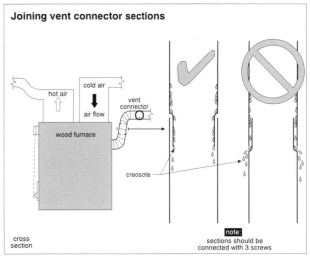

hot air

cold air
air flow

vent connector

wood furnace

creosote

cross section

note:
sections should be
connected with 3 screws

0 2 5 1

Vent connector support and slope

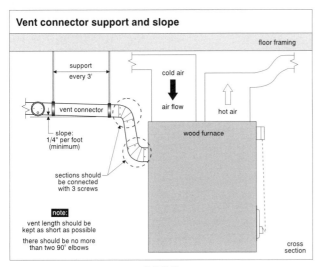

floor framing

support
every 3'

cold air
air flow

hot air

vent connector

wood furnace

slope:
1/4" per foot
(minimum)

sections should
be connected
with 3 screws

note:
vent length should be
kept as short as possible

there should be no more
than two 90° elbows

cross
section

0 2 5 2

Thermostat conditions to watch for

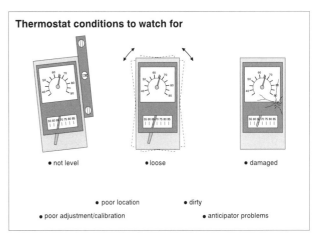

● not level

● loose

● damaged

● poor location

● dirty

● poor adjustment/calibration

● anticipator problems

0 2 5 3

Fan/limit switch

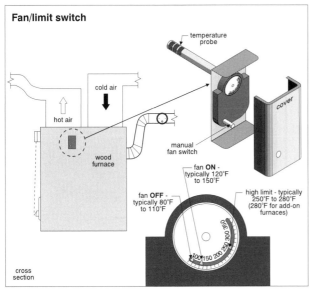

temperature
probe

cold air

hot air

cover

wood
furnace

manual
fan switch

fan **ON** -
typically 120°F
to 150°F

fan **OFF** -
typically 80°F
to 110°F

high limit - typically
250°F to 280°F
(280°F for add-on
furnaces)

cross
section

0 2 5 4

Duct clearances - typical

floor framing

2"

6"

6"

cold air

plenum

hot air

6 feet

wood furnace

cross
section

0 2 5 5

Combustible return ducts

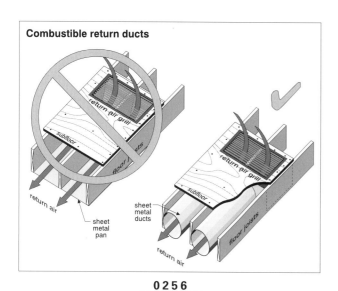

return air

sheet
metal
pan

return air grill

subfloor

floor joists

return air grill

subfloor

sheet
metal
ducts

floor joists

return air

0256

Add-on wood furnace
(downstream series)

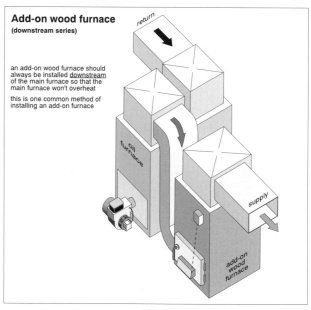

return

an add-on wood furnace should
always be installed <u>downstream</u>
of the main furnace so that the
main furnace won't overheat

this is one common method of
installing an add-on furnace

oil
furnace

supply

add-on
wood
furnace

0257

Add-on wood furnace
(divider plate method)

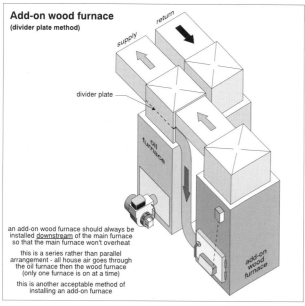

supply

return

divider plate

oil
furnace

add-on
wood
furnace

an add-on wood furnace should always be
installed <u>downstream</u> of the main furnace
so that the main furnace won't overheat

this is a series rather than parallel
arrangement - all house air goes through
the oil furnace then the wood furnace
(only one furnace is on at a time)

this is another acceptable method of
installing an add-on furnace

0258

Wood furnace not downstream

with the add-on wood furnace located
upstream, the main (oil furnace) blower,
motor, belts, filters, etc. could overheat

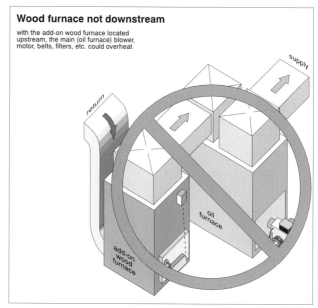

supply

return

oil
furnace

add-on
wood
furnace

0259

Duct arrangement allows reverse flow

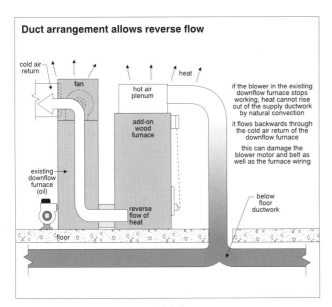

cold air return

fan

heat

hot air plenum

add-on wood furnace

if the blower in the existing downflow furnace stops working, heat cannot rise out of the supply ductwork by natural convection

it flows backwards through the cold air return of the downflow furnace

this can damage the blower motor and belt as well as the furnace wiring

existing downflow furnace (oil)

reverse flow of heat

below floor ductwork

floor

0260

Dual fuel furnace (wood/oil)

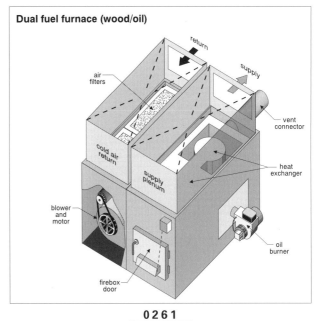

return

supply

air filters

vent connector

cold air return

supply plenum

heat exchanger

blower and motor

oil burner

firebox door

0261

Wood stove components

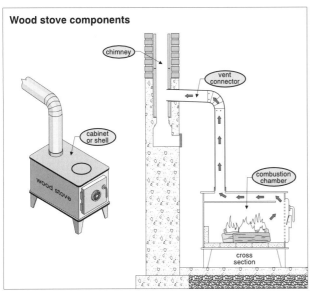

chimney

vent connector

cabinet or shell

combustion chamber

wood stove

cross section

0262

Don't confine wood stoves

wood stoves should not be located in small, confined spaces as they may be starved for combustion air and/or overheat

cross section

wood stove

0263

Chimney liners often required for fireplace inserts

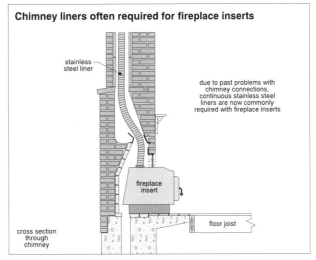

stainless steel liner

due to past problems with chimney connections, continuous stainless steel liners are now commonly required with fireplace inserts

fireplace insert

floor joist

cross section through chimney

0264

Convective wood stoves

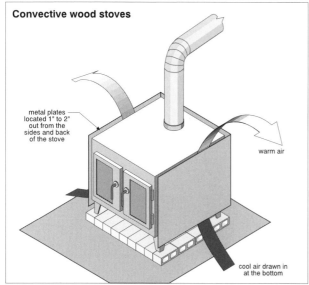

metal plates located 1" to 2" out from the sides and back of the stove

warm air

cool air drawn in at the bottom

0265

Advanced combustion wood stoves

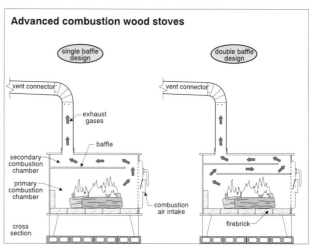

single baffle design

double baffle design

vent connector

exhaust gases

baffle

secondary combustion chamber

primary combustion chamber

combustion air intake

cross section

firebrick

0266

Catalytic combustors

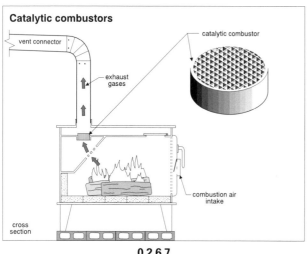

catalytic combustor

vent connector

exhaust gases

combustion air intake

cross section

0267

Bypass damper on catalytic wood stove

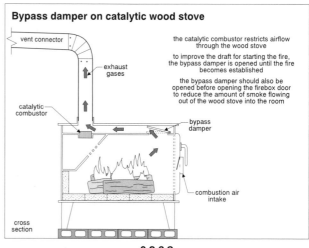

the catalytic combustor restricts airflow through the wood stove

to improve the draft for starting the fire, the bypass damper is opened until the fire becomes established

the bypass damper should also be opened before opening the firebox door to reduce the amount of smoke flowing out of the wood stove into the room

vent connector

exhaust gases

catalytic combustor

bypass damper

combustion air intake

cross section

0268

Pellet stoves

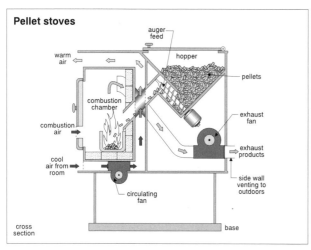

auger feed

warm air

hopper

pellets

combustion chamber

exhaust fan

combustion air

exhaust products

cool air from room

side wall venting to outdoors

circulating fan

cross section

base

0269

Wood stove floor protection - masonry
(legs less than 3" tall)

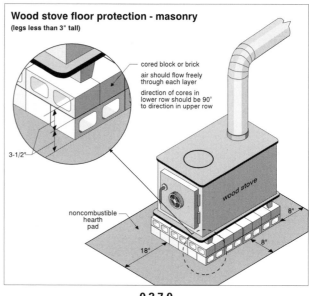

cored block or brick

air should flow freely through each layer

direction of cores in lower row should be 90° to direction in upper row

3-1/2"

noncombustible hearth pad

wood stove

8"

8"

18"

0270

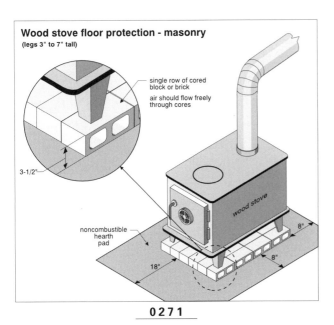

Wood stove floor protection - masonry
(legs 3" to 7" tall)

single row of cored block or brick

air should flow freely through cores

3-1/2"

noncombustible hearth pad

8"

8"

18"

0271

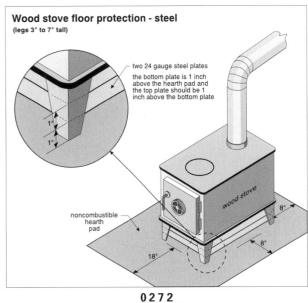

Wood stove floor protection - steel
(legs 3" to 7" tall)

two 24 gauge steel plates

the bottom plate is 1 inch above the hearth pad and the top plate should be 1 inch above the bottom plate

1"

1"

noncombustible hearth pad

8"

18"

8"

0272

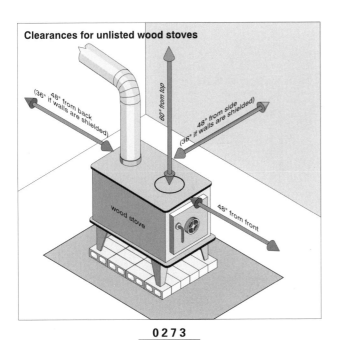

Clearances for unlisted wood stoves

(36" if walls are shielded)
48" from back

60" from top

48" from side
(36" if walls are shielded)

48" from front

0273

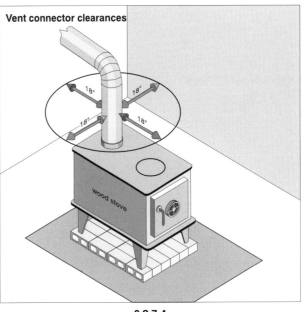

Vent connector clearances

18"   18"

18"   18"

0274

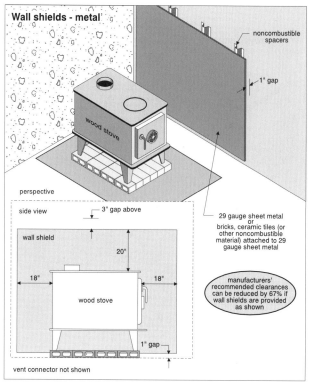

Wall shields - metal

noncombustible spacers

1" gap

wood stove

perspective

side view

3" gap above

wall shield

20"

18"   18"

wood stove

1" gap

vent connector not shown

29 gauge sheet metal
or
bricks, ceramic tiles (or
other noncombustible
material) attached to 29
gauge sheet metal

manufacturers'
recommended clearances
can be reduced by 67% if
wall shields are provided
as shown

0275

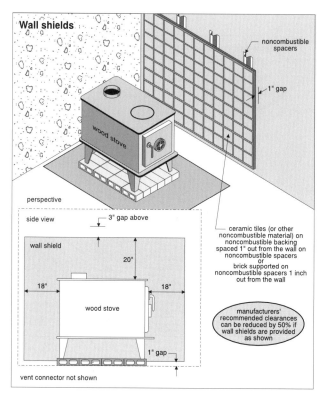

Wall shields

noncombustible spacers

1" gap

wood stove

perspective

side view

3" gap above

wall shield

20"

18"   18"

wood stove

1" gap

vent connector not shown

ceramic tiles (or other
noncombustible material) on
noncombustible backing
spaced 1" out from the wall on
noncombustible spacers
or
brick supported on
noncombustible spacers 1 inch
out from the wall

manufacturers'
recommended clearances
can be reduced by 50% if
wall shields are provided
as shown

0276

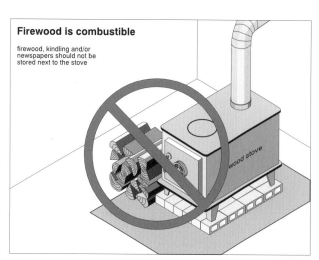

Firewood is combustible

firewood, kindling and/or
newspapers should not be
stored next to the stove

wood stove

0277

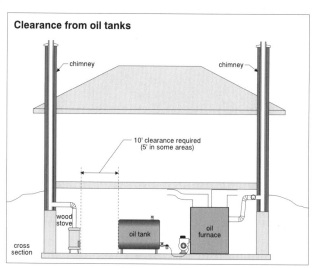

Clearance from oil tanks

chimney   chimney

10' clearance required
(5' in some areas)

wood
stove

oil tank   oil
furnace

cross
section

0278

Wood stove vent connector slope

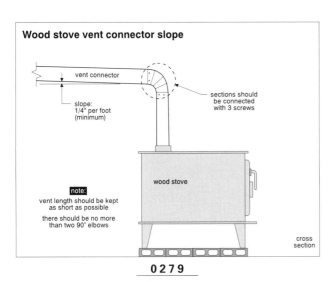

vent connector

slope:
1/4" per foot
(minimum)

sections should
be connected
with 3 screws

wood stove

note:
vent length should be kept
as short as possible

there should be no more
than two 90° elbows

cross
section

0279

Wood stove vent connections

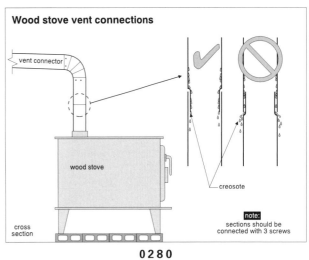

vent connector

wood stove

creosote

cross
section

note:
sections should be
connected with 3 screws

0280

Draw bands or adjustable connectors

are used to allow for expansion in long, straight flue sections

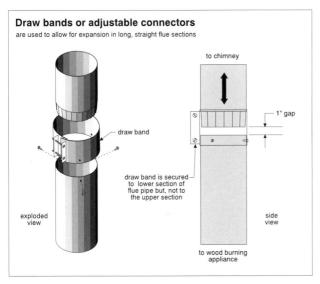

to chimney

draw band

1" gap

draw band is secured
to lower section of
flue pipe but, not to
the upper section

exploded
view

side
view

to wood burning
appliance

0281

Poor vent connection at chimney

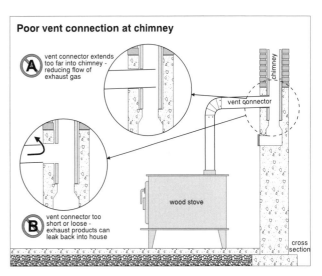

A vent connector extends
too far into chimney -
reducing flow of
exhaust gas

chimney

vent connector

B vent connector too
short or loose -
exhaust products can
leak back into house

wood stove

cross
section

0282

Testing dampers

unless the wood stove is operating, you will want to see if the damper is operational

remember to return the damper to its original position

convective wood stove

0283

What we'll be looking at:

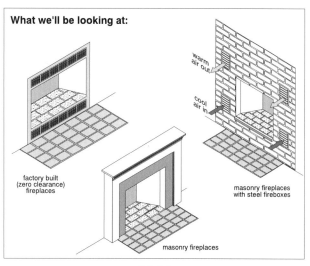

warm air out

cool air in

factory built (zero clearance) fireplaces

masonry fireplaces with steel fireboxes

masonry fireplaces

0284

Masonry fireplace components

flue liner

chimney cap

chimney

chimney walls

roof rafter

ceiling joist

drywall or plaster

chimney

smoke chamber

mantel shelf

damper

throat

smoke shelf

hearth

firebox

hearth extension

floor joist

ash dump

ash pit

basement

chimney cleanout

foundation wall

cross section through chimney

footing

basement floor slab

0285

Combustible clearance requirements

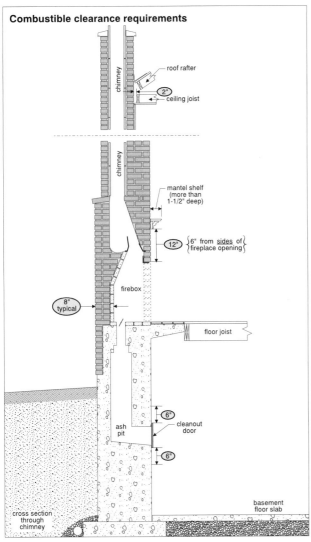

roof rafter

2"

ceiling joist

chimney

chimney

mantel shelf
(more than
1-1/2" deep)

12" { 6" from *sides* of
fireplace opening }

firebox

8"
typical

floor joist

ash
pit

cleanout
door

6"

6"

basement
floor slab

cross section
through
chimney

0286

Basement fireplaces

this illustration shows some of the
reasons why basement fireplaces
often have poor draft

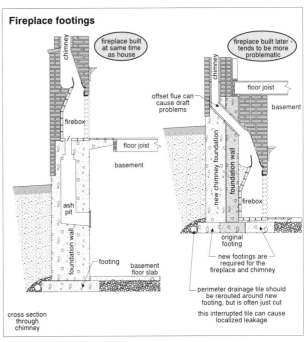

D chimney height and
stack effect promote
downdrafts

chimney

flue

B number of
chimney flue
offsets

first floor

basement

A lack of combustion air
due to competition
with other appliances
and negative pressure
commonly found in
basements

C low temperatures
at the back wall
of the firebox

0287

Problematic fireplace designs

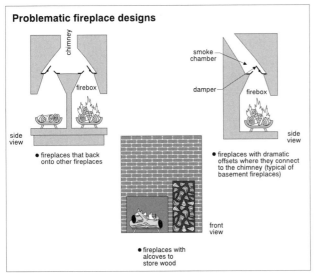

chimney

smoke
chamber

firebox

damper

firebox

side
view

side
view

● fireplaces that back
onto other fireplaces

● fireplaces with dramatic
offsets where they connect
to the chimney (typical of
basement fireplaces)

front
view

● fireplaces with
alcoves to
store wood

0288

Fireplace footings

fireplace built
at same time
as house

fireplace built later -
tends to be more
problematic

chimney

chimney

offset flue can
cause draft
problems

floor joist

basement

firebox

floor joist

new chimney foundation

foundation wall

basement

ash
pit

firebox

foundation wall

footing

original
footing

basement
floor slab

new footings are
required for the
fireplace and chimney

perimeter drainage tile should
be rerouted around new
footing, but is often just cut

this interrupted tile can cause
localized leakage

cross section
through
chimney

0289

Causes of chimney settling or leaning

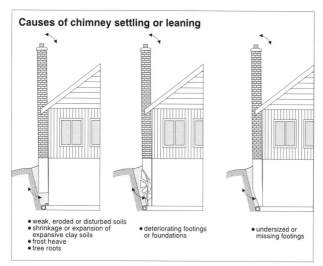

- weak, eroded or disturbed soils
- shrinkage or expansion of expansive clay soils
- frost heave
- tree roots

- deteriorating footings or foundations

- undersized or missing footings

0290

Hearth materials

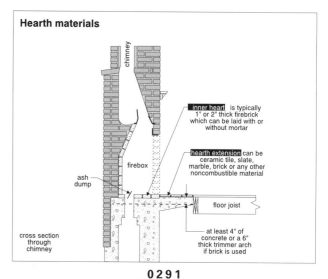

chimney

inner hearth is typically 1" or 2" thick firebrick which can be laid with or without mortar

firebox

hearth extension can be ceramic tile, slate, marble, brick or any other noncombustible material

ash dump

floor joist

cross section through chimney

at least 4" of concrete or a 6" thick trimmer arch if brick is used

0291

Metal spark strip

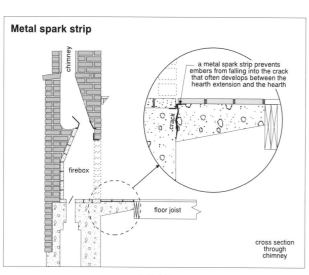

chimney

a metal spark strip prevents embers from falling into the crack that often develops between the hearth extension and the hearth

crack

firebox

floor joist

cross section through chimney

0292

Hearth extension dimensions

firebox

16" to 20"

8" to 12"

hearth

0293

Raised hearths

the hearth should be extended out to 20" when the firebox is elevated

in some areas, the hearth depth must be increased in stages depending on how far above the floor the firebox is located

the area between the firebox and the hearth should be noncombustible

20"

8"

hearth

0294

Hearth extensions for factory built fireplaces

factory built fireplace

metal chimney

wall framing

warm air out

factory built fireplace

perspective view

metal spark strip

note: hearth extension can be laid directly over a wood floor

hearth extension

cool air in

subfloor

metal spark strip

floor joists

cross section

0295

Remove hearth extension forming boards

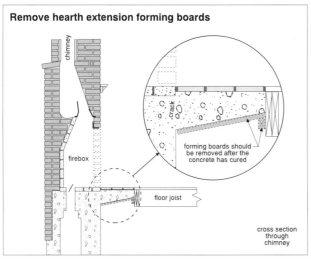

chimney

firebox

crack

forming boards should be removed after the concrete has cured

floor joist

cross section through chimney

0296

Firebrick on walls

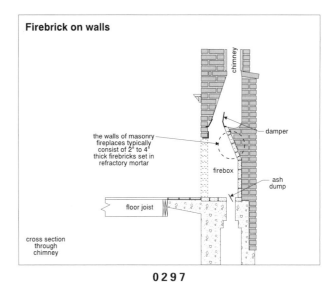

chimney

the walls of masonry fireplaces typically consist of 2" to 4" thick firebricks set in refractory mortar

damper

firebox

ash dump

floor joist

cross section through chimney

0297

Metal firebox clearance

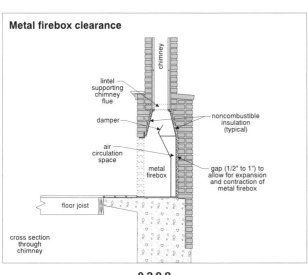

chimney

lintel supporting chimney flue

damper

air circulation space

metal firebox

noncombustible insulation (typical)

gap (1/2" to 1") to allow for expansion and contraction of metal firebox

floor joist

cross section through chimney

0298

Flue liner shouldn't rest on metal firebox

leave gap here to allow for expansion of firebox

clay flue liner

lintel supporting chimney flue

flexible filler

top of metal firebox

gap (1/2" to 1") filled with noncombustible insulation

chimney

metal firebox

floor joist

cross section through chimney

0299

Coal burning fireplaces

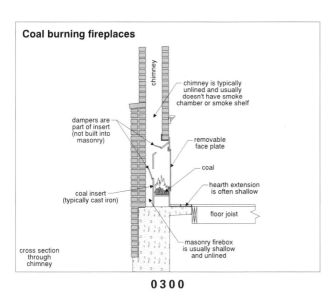

chimney

chimney is typically unlined and usually doesn't have smoke chamber or smoke shelf

dampers are part of insert (not built into masonry)

removable face plate

coal

coal insert (typically cast iron)

hearth extension is often shallow

floor joist

masonry firebox is usually shallow and unlined

cross section through chimney

0300

Lintel rusting, sagging or loose

chimney

rusting due to water leakage from above or lintel undersizing can cause cracking of the fireplace face

crack

lintel

firebox

firebox

floor joist

perspective

cross section through chimney

0301

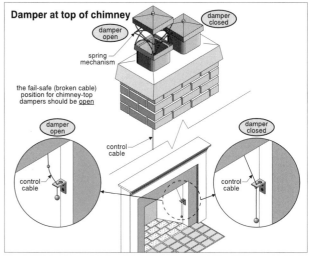

Firebox damper

damper

smoke shelf

damper handle

lintel

chimney

firebox

floor joist

damper handles may be located within the firebox or there may be a handle on the fireplace face

cross section through chimney

0302

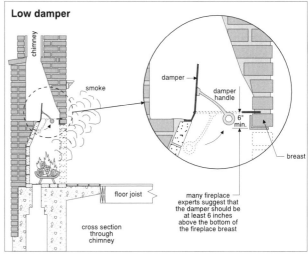

Damper at top of chimney

damper open

damper closed

spring mechanism

the fail-safe (broken cable) position for chimney-top dampers should be open

damper open

control cable

damper closed

control cable

control cable

0303

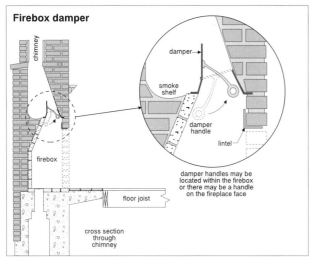

Missing damper

chimney

dampers must sometimes be removed to install fireplace inserts

if the insert is removed later, the fireplace will be without a damper - resulting in heat loss and possible downdraft problems

damper may have been removed

fireplace insert

cross section through chimney

floor joist

0304

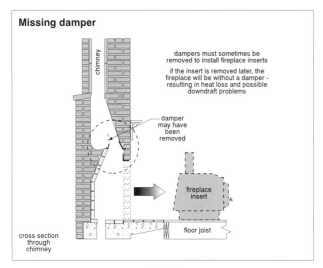

Low damper

chimney

smoke

damper

damper handle

6" min.

breast

floor joist

many fireplace experts suggest that the damper should be at least 6 inches above the bottom of the fireplace breast

cross section through chimney

0305

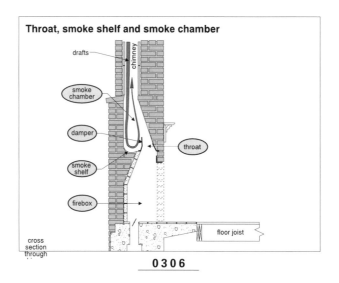

Throat, smoke shelf and smoke chamber

drafts

chimney

smoke chamber

damper

throat

smoke shelf

firebox

floor joist

cross section through

0306

Smoke chamber - wall slope

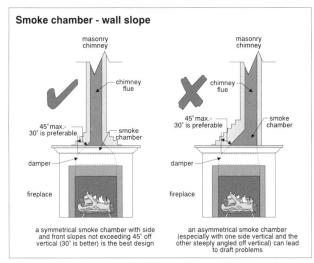

masonry chimney

chimney flue

45° max.- 30° is preferable

smoke chamber

damper

fireplace

a symmetrical smoke chamber with side and front slopes not exceeding 45° off vertical (30° is better) is the best design

masonry chimney

chimney flue

smoke chamber

45° max.- 30° is preferable

damper

fireplace

an asymmetrical smoke chamber (especially with one side vertical and the other steeply angled off vertical) can lead to draft problems

0 3 0 7

Smoke chamber forms must be removed

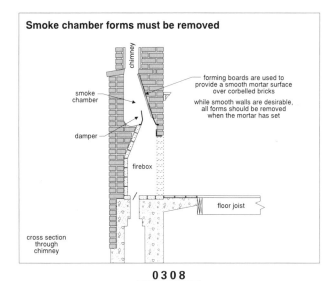

chimney

smoke chamber

damper

firebox

forming boards are used to provide a smooth mortar surface over corbelled bricks

while smooth walls are desirable, all forms should be removed when the mortar has set

floor joist

cross section through chimney

0 3 0 8

Fireplace breast

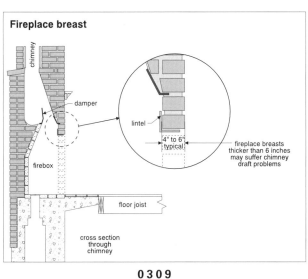

chimney

damper

lintel

4" to 6" typical

fireplace breasts thicker than 6 inches may suffer chimney draft problems

firebox

floor joist

cross section through chimney

0 3 0 9

Settled (gap at wall)

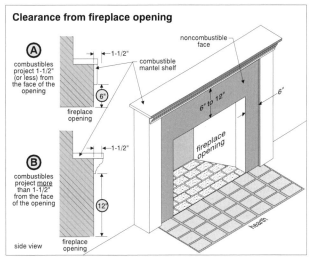

chimney

gap develops - watch for smoke staining on the wall behind the mantel shelf

rotation

mantel shelf

stone veneer fireplace face

firebox

floor joist

downward

sag

cross section through chimney

0 3 1 0

Clearance from fireplace opening

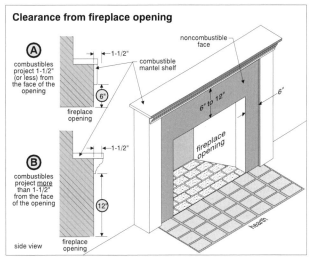

Ⓐ combustibles project 1-1/2" (or less) from the face of the opening

1-1/2"

combustible mantel shelf

noncombustible face

6"

fireplace opening

6" to 12"

6"

fireplace opening

Ⓑ combustibles project more than 1-1/2" from the face of the opening

1-1/2"

12"

side view

fireplace opening

hearth

0 3 1 1

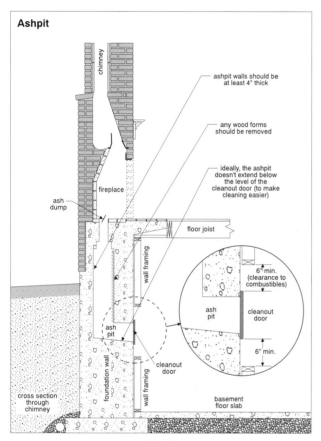

Ashpit

chimney

ashpit walls should be at least 4" thick

any wood forms should be removed

ideally, the ashpit doesn't extend below the level of the cleanout door (to make cleaning easier)

fireplace

ash dump

floor joist

wall framing

6" min. (clearance to combustibles)

ash pit

cleanout door

ash pit

6" min.

foundation wall

wall framing

cleanout door

cross section through chimney

basement floor slab

0 3 1 2

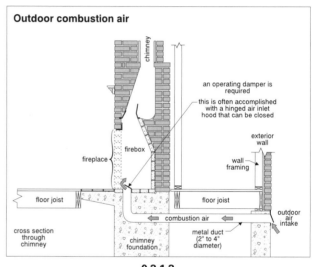

Outdoor combustion air

chimney

an operating damper is required

this is often accomplished with a hinged air inlet hood that can be closed

firebox

fireplace

exterior wall

wall framing

floor joist

floor joist

cross section through chimney

combustion air

outdoor air intake

chimney foundation

metal duct (2" to 4" diameter)

0 3 1 3

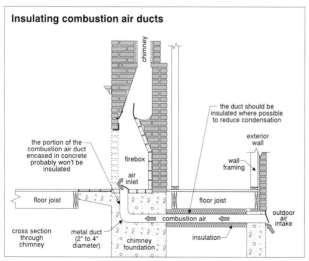

Insulating combustion air ducts

chimney

the duct should be insulated where possible to reduce condensation

the portion of the combustion air duct encased in concrete probably won't be insulated

firebox

exterior wall

air inlet

wall framing

floor joist

floor joist

cross section through chimney

metal duct (2" to 4" diameter)

combustion air

outdoor air intake

chimney foundation

insulation

0 3 1 4

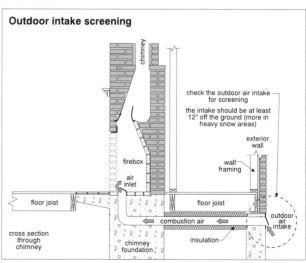

Outdoor intake screening

chimney

check the outdoor air intake for screening

the intake should be at least 12" off the ground (more in heavy snow areas)

firebox

exterior wall

air inlet

wall framing

floor joist

floor joist

cross section through chimney

combustion air

outdoor air intake

chimney foundation

insulation

0 3 1 5

Combustion air duct clearances

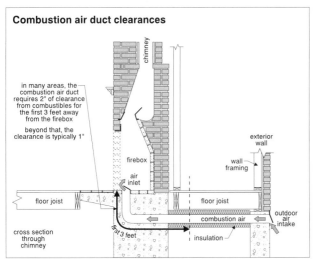

chimney

in many areas, the combustion air duct requires 2" of clearance from combustibles for the first 3 feet away from the firebox

beyond that, the clearance is typically 1"

exterior wall

firebox

wall framing

air inlet

floor joist

floor joist

outdoor air intake

combustion air

cross section through chimney

first 3 feet

insulation

0 3 1 6

Glass doors

glass doors decrease the amount of heat radiated into the room by the fireplace but, also reduce the amount of warm household air that goes up the chimney

things to watch for:

frame rusty or warped

inoperative

cracked or broken glass

0 3 1 7

Heat circulators

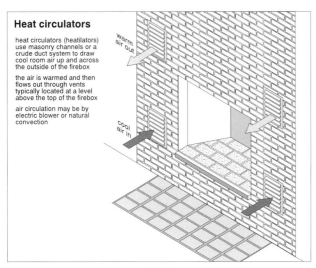

heat circulators (heatilators) use masonry channels or a crude duct system to draw cool room air up and across the outside of the firebox

the air is warmed and then flows out through vents typically located at a level above the top of the firebox

air circulation may be by electric blower or natural convection

warm air out

cool air in

0 3 1 8

Gas igniters

most gas igniters have few safety features and may not be permitted in some areas - find out if they are allowed in <u>your</u> area

wood burning fireplace

key

(A) open the gas valve

(B) ignite the burner with a match or burning paper

the burner pipe should not be buried in ash or embers

burner

adjustable air shutter

manual operator valve (outside the firebox)

gas supply

0 3 1 9

How steam systems work

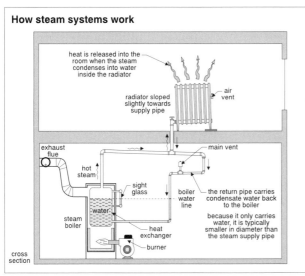

heat is released into the room when the steam condenses into water inside the radiator

radiator sloped slightly towards supply pipe

air vent

exhaust flue

hot steam

main vent

sight glass

boiler water line

the return pipe carries condensate water back to the boiler

because it only carries water, it is typically smaller in diameter than the steam supply pipe

steam boiler

water

heat exchanger

burner

cross section

0 3 2 0

Steam system operation: at rest

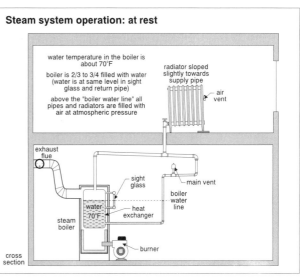

water temperature in the boiler is about 70°F

boiler is 2/3 to 3/4 filled with water (water is at same level in sight glass and return pipe)

above the "boiler water line" all pipes and radiators are filled with air at atmospheric pressure

radiator sloped slightly towards supply pipe

air vent

exhaust flue

sight glass

main vent

boiler water line

steam boiler

water 70°F

heat exchanger

burner

cross section

0 3 2 1

Steam system operation: call for heat

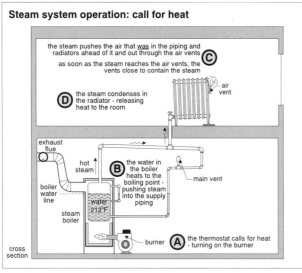

the steam pushes the air that <u>was</u> in the piping and radiators ahead of it and out through the air vents **C**

as soon as the steam reaches the air vents, the vents close to contain the steam

D the steam condenses in the radiator - releasing heat to the room

air vent

exhaust flue

hot steam

B the water in the boiler heats to the boiling point - pushing steam into the supply piping

main vent

boiler water line

water 212°F

steam boiler

burner **A** the thermostat calls for heat - turning on the burner

cross section

0322

Air vents

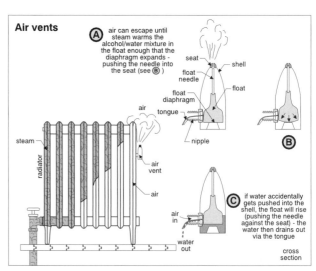

A air can escape until steam warms the alcohol/water mixture in the float enough that the diaphragm expands - pushing the needle into the seat (see **B**)

seat
shell
float needle
float
float diaphragm
tongue
nipple

steam
radiator
air
air vent
air
air in
water out

B

C if water accidentally gets pushed into the shell, the float will rise (pushing the needle against the seat) - the water then drains out via the tongue

cross section

0323

Dimension "A"

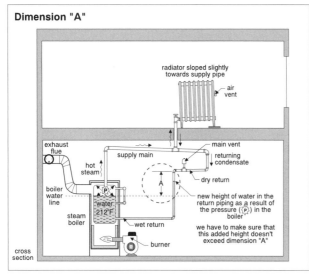

radiator sloped slightly towards supply pipe

air vent

exhaust flue

supply main

hot steam

main vent

returning condensate

dry return

boiler water line

water 212°F

steam boiler

A

new height of water in the return piping as a result of the pressure (P) in the boiler

wet return

we have to make sure that this added height doesn't exceed dimension "A"

burner

cross section

0324

Equalizer pipe

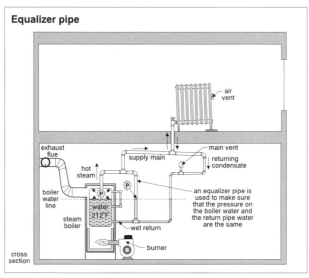

air vent

exhaust flue

supply main

hot steam

main vent

returning condensate

boiler water line

water 212°F

steam boiler

an equalizer pipe is used to make sure that the pressure on the boiler water and the return pipe water are the same

wet return

burner

cross section

0325

Steam system operation: thermostat is satisfied

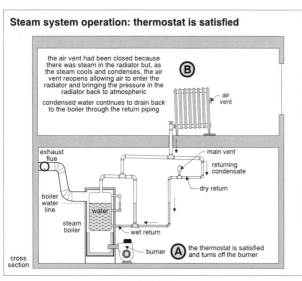

the air vent had been closed because there was steam in the radiator but, as the steam cools and condenses, the air vent reopens allowing air to enter the radiator and bringing the pressure in the radiator back to atmospheric **B**

condensed water continues to drain back to the boiler through the return piping

air vent

exhaust flue

main vent

returning condensate

dry return

boiler water line

water

steam boiler

wet return

burner **A** the thermostat is satisfied and turns off the burner

cross section

0326

Hartford Loop

air vent

exhaust flue

supply main

hot steam

main vent

returning condensate

dry return

boiler water line

water

steam boiler

the Hartford Loop prevents water from draining out of the boiler if there is a leak in the wet return lines (upstream of the loop)

burner

cross section

0327

One pipe counterflow system

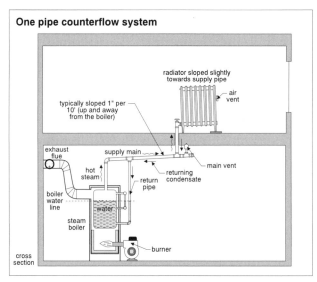

0 3 2 8

One pipe parallel flow system

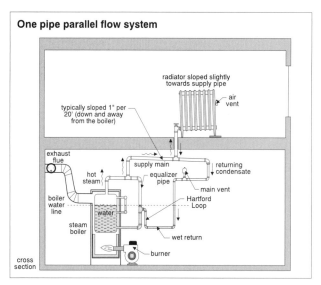

0 3 2 9

Two pipe system

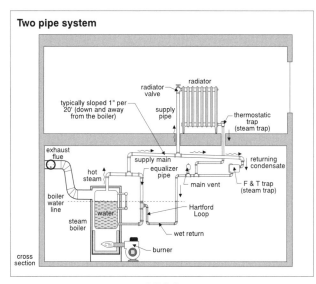

0 3 3 0

Thermostatic trap

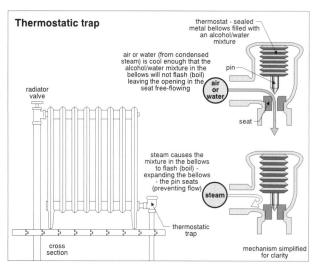

0 3 3 1

F & T trap (float and thermostat trap)

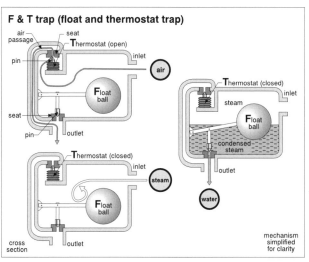

0 3 3 2

Location of condensate pump

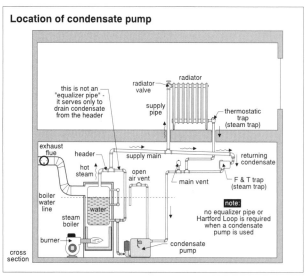

0 3 3 3

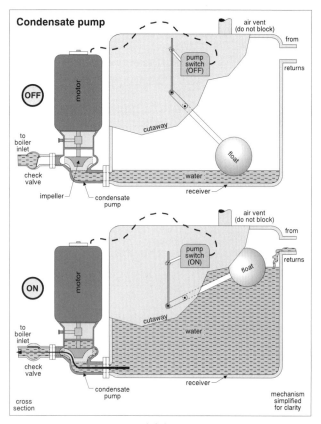

Condensate pump

air vent
(do not block)

from

returns

pump
switch
(OFF)

(OFF)

motor

cutaway

float

to
boiler
inlet

check
valve

water

impeller

condensate
pump

receiver

air vent
(do not block)

from

returns

pump
switch
(ON)

float

(ON)

motor

cutaway

water

to
boiler
inlet

check
valve

condensate
pump

receiver

mechanism
simplified
for clarity

cross
section

0334

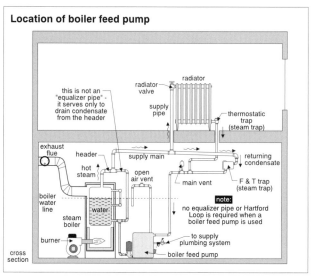

Location of boiler feed pump

this is not an
"equalizer pipe" -
it serves only to
drain condensate
from the header

radiator

radiator
valve

supply
pipe

thermostatic
trap
(steam trap)

exhaust
flue

header

supply main

returning
condensate

hot
steam

open
air vent

main vent

F & T trap
(steam trap)

boiler
water
line

water

note:
no equalizer pipe or Hartford
Loop is required when a
boiler feed pump is used

steam
boiler

to supply
plumbing system

burner

boiler feed pump

cross
section

0335

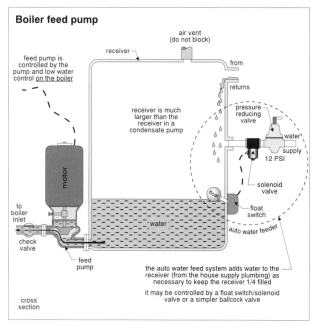

Boiler feed pump

air vent
(do not block)

receiver

from

feed pump is
controlled by the
pump and low water
control on the boiler

returns

receiver is much
larger than the
receiver in a
condensate pump

pressure
reducing
valve

water
supply
12 PSI

motor

solenoid
valve

to
boiler
inlet

float

check
valve

water

float
switch

auto water feeder

feed
pump

the auto water feed system adds water to the
receiver (from the house supply plumbing) as
necessary to keep the receiver 1/4 filled

cross
section

it may be controlled by a float switch/solenoid
valve or a simpler ballcock valve

0336

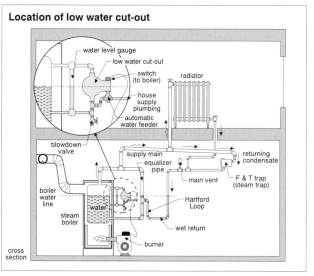

Location of low water cut-out

water level gauge

low water cut-out

switch
(to boiler)

radiator

house
supply
plumbing

automatic
water feeder

blowdown
valve

supply main

returning
condensate

equalizer
pipe

main vent

F & T trap
(steam trap)

boiler
water
line

water

Hartford
Loop

steam
boiler

wet return

burner

cross
section

0337

Pressure relief valve

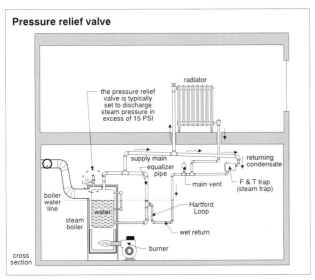

the pressure relief valve is typically set to discharge steam pressure in excess of 15 PSI

radiator

supply main

equalizer pipe

returning condensate

main vent

F & T trap (steam trap)

boiler water line

Hartford Loop

water

steam boiler

wet return

burner

cross section

0338

Location of pressuretrol

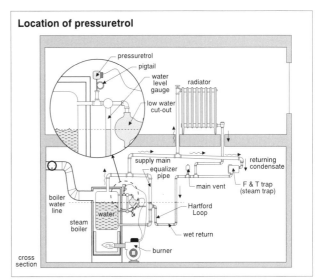

pressuretrol

pigtail

water level gauge

low water cut-out

radiator

supply main

equalizer pipe

returning condensate

main vent

F & T trap (steam trap)

boiler water line

Hartford Loop

water

steam boiler

wet return

burner

cross section

0339

Hartford Loop and equalizer

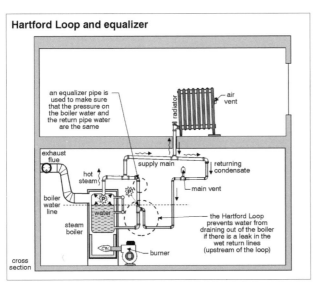

an equalizer pipe is used to make sure that the pressure on the boiler water and the return pipe water are the same

air vent

radiator

exhaust flue

hot steam

supply main

returning condensate

main vent

boiler water line

water

steam boiler

the Hartford Loop prevents water from draining out of the boiler if there is a leak in the wet return lines (upstream of the loop)

burner

cross section

0340

Water hammer

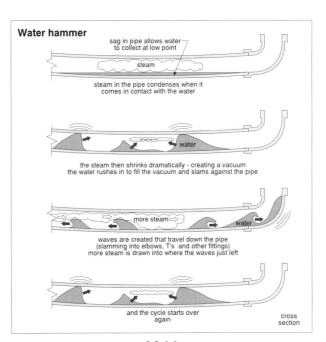

sag in pipe allows water to collect at low point

steam

steam in the pipe condenses when it comes in contact with the water

steam

water

the steam then shrinks dramatically - creating a vacuum the water rushes in to fill the vacuum and slams against the pipe

more steam

water

waves are created that travel down the pipe (slamming into elbows, T's and other fittings) more steam is drawn into where the waves just left

and the cycle starts over again

cross section

0341

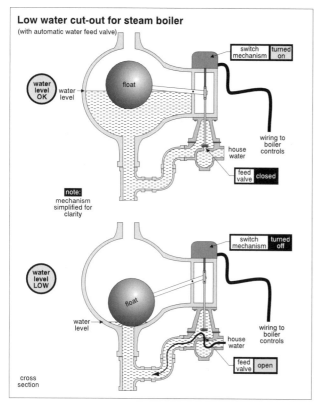

Low water cut-out for steam boiler
(with automatic water feed valve)

water level OK

float

water level

switch mechanism — turned on

wiring to boiler controls

house water

feed valve — closed

note: mechanism simplified for clarity

water level LOW

float

water level

switch mechanism — turned off

wiring to boiler controls

house water

feed valve — open

cross section

0342

Water level gauge

the water level gauge (sight gauge, sight glass) should be installed so that the boiler water line is about midway up the gauge when the boiler is at rest

don't fire up the boiler if you don't see water in the gauge

radiator

supply main

equalizer pipe

returning condensate

main vent

F & T trap (steam trap)

boiler water line

water

Hartford Loop

steam boiler

wet return

burner

cross section

0343

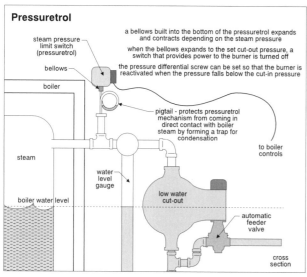

Pressuretrol

a bellows built into the bottom of the pressuretrol expands and contracts depending on the steam pressure

when the bellows expands to the set cut-out pressure, a switch that provides power to the burner is turned off

the pressure differential screw can be set so that the burner is reactivated when the pressure falls below the cut-in pressure

steam pressure limit switch (pressuretrol)

bellows

boiler

pigtail - protects pressuretrol mechanism from coming in direct contact with boiler steam by forming a trap for condensation

to boiler controls

steam

water level gauge

boiler water level

low water cut-out

automatic feeder valve

cross section

0344

Hartford Loop and equalizer - details

boiler water line

2" to 3"

diameter of equalizer pipe should be at least 2"

Hartford Loop

radiator

air vent

exhaust flue

hot steam

supply main

returning condensate

main vent

boiler water line

water

steam boiler

burner

cross section

0345

Hartford Loop - close nipple missing

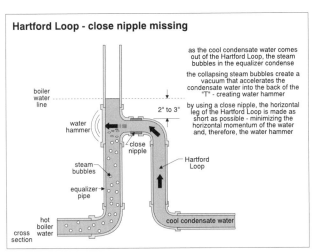

as the cool condensate water comes out of the Hartford Loop, the steam bubbles in the equalizer condense

the collapsing steam bubbles create a vacuum that accelerates the condensate water into the back of the "T" - creating water hammer

by using a close nipple, the horizontal leg of the Hartford Loop is made as short as possible - minimizing the horizontal momentum of the water and, therefore, the water hammer

boiler water line

2" to 3"

water hammer

close nipple

steam bubbles

equalizer pipe

Hartford Loop

hot boiler water

cool condensate water

cross section

0 3 4 6

Main air vent

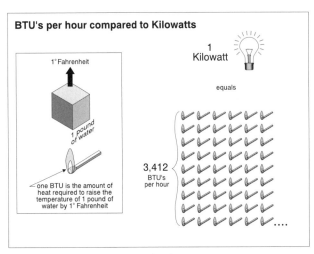

the main vent works very similarly to the air vents found on some radiators - air can escape through the vent, but steam can't escape

air can also be brought back into the system through the air vent when the steam has gone

6" to 10"

main vent

15"

supply pipe

thermostatic trap (steam trap)

exhaust flue

supply main

returning condensate

hot steam

main vent

F & T trap (steam trap)

boiler water line

water

Hartford Loop

wet return

steam boiler

burner

cross section

0 3 4 7

Current flow = amperage

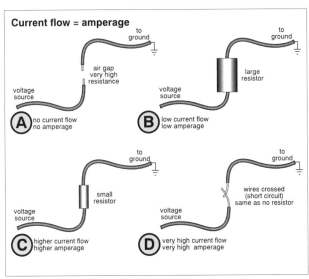

to ground

to ground

air gap very high resistance

large resistor

voltage source

voltage source

(A) no current flow no amperage

(B) low current flow low amperage

to ground

to ground

small resistor

wires crossed (short circuit) same as no resistor

voltage source

voltage source

(C) higher current flow higher amperage

(D) very high current flow very high amperage

0 3 4 8

BTU's per hour compared to Kilowatts

1° Fahrenheit

1 pound of water

one BTU is the amount of heat required to raise the temperature of 1 pound of water by 1° Fahrenheit

1 Kilowatt

equals

3,412 BTU's per hour

....

0 3 4 9

Equivalent furnaces

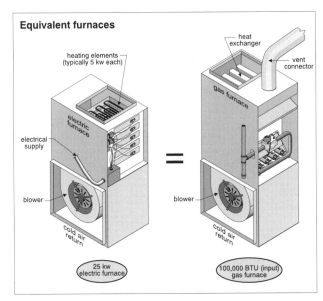

heating elements (typically 5 kw each)

heat exchanger

vent connector

gas furnace

electric furnace

electrical supply

blower

blower

cold air return

cold air return

=

25 kw electric furnace

100,000 BTU (input) gas furnace

0 3 5 0

5 to 8 watts per square foot

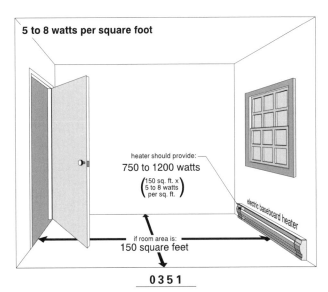

heater should provide:

750 to 1200 watts

(150 sq. ft. x 5 to 8 watts per sq. ft.)

electric baseboard heater

if room area is:

150 square feet

0 3 5 1

Wire, fuse and heater sizing

In some areas, a fuse or breaker on an electric heater circuit may be larger than the wire ampacity but still be acceptable

electric heater load can be 100% of wire ampacity (15 amps X 240 volts) equals 3600 watts

Ⓑ

Ⓒ fuse can be 125% of the wire or heater load (15 amps X 125%) - roughly 20 amps

14 gauge wire

Ⓐ wire ampacity is 15 amps

20 AMP FUSE
20 AMP FUSE

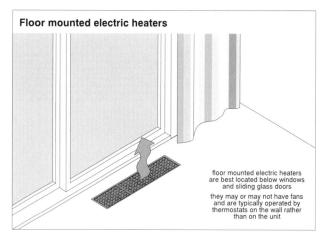

0352

Floor mounted electric heaters

floor mounted electric heaters are best located below windows and sliding glass doors

they may or may not have fans and are typically operated by thermostats on the wall rather than on the unit

0353

Toekick electric heaters

toekick heaters are often found in renovated kitchens

these typically use fans

check carefully for one of these before reporting that a kitchen has no source of heat

toekick
air

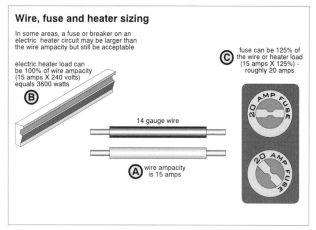

0354

Electric wall heater - fan operated

warm air out

thermostat

cool air in

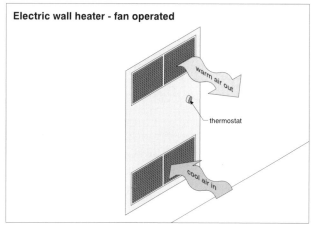

0355

Best location for electric heaters

the coolest spots in a room tend to be at floor level by outside walls (particularly below windows) so, electric heaters are best located in these areas

cool air

hot air

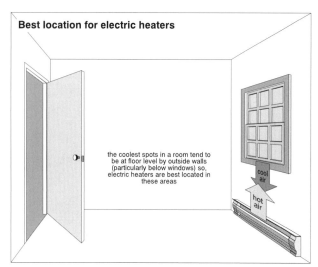

0356

Wiring for electric space heaters

distribution panel

from service box

to electric heater

STOVE

grounding terminal

ground wire

to electric heater

if conventional wiring is used for electric heater circuits, the white wire should be wrapped with black electrical tape to show that it is a hot wire

wiring intended specifically for electric heater circuits typically has red or orange sheathing and contains one black and one red wire

black wire

red wire

the fuses/breakers for 240 volt electric heaters must be linked together

wrap white wire with black tape

neutral bus bar

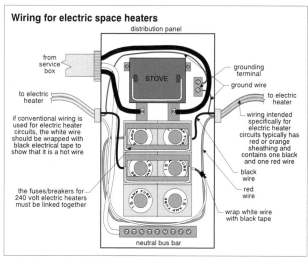

0357

Determining baseboard heater wattage - rule of thumb

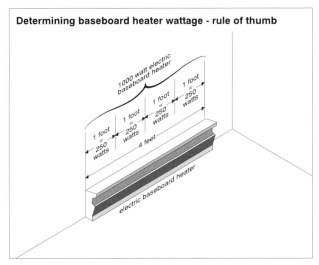

1000 watt electric baseboard heater

1 foot = 250 watts
1 foot = 250 watts
1 foot = 250 watts
1 foot = 250 watts

4 feet

electric baseboard heater

0358

Electric baseboard heaters - clearances to draperies

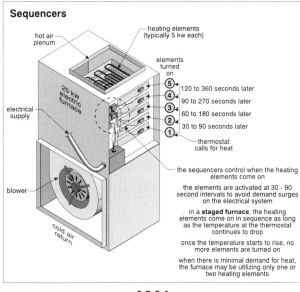

OR

8" above

1"

3" in front and
1" above the floor

0359

Outlets should not be above electric baseboard heaters

cords plugged into outlets above electric baseboard heaters could overheat if accidentally draped over the heater

outlets should be located at either end of the heater

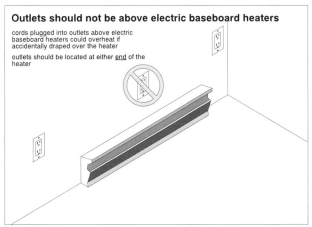

0360

Sequencers

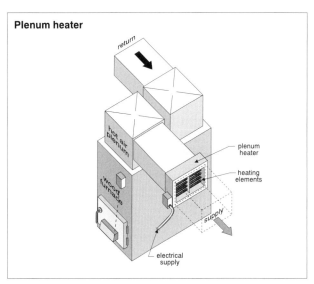

hot air plenum

heating elements (typically 5 kw each)

elements turned on

25 kw electric furnace

⑤ 120 to 360 seconds later
④ 90 to 270 seconds later
③ 60 to 180 seconds later
② 30 to 90 seconds later
① thermostat calls for heat

electrical supply

blower

cold air return

the sequencers control when the heating elements come on

the elements are activated at 30 - 90 second intervals to avoid demand surges on the electrical system

in a **staged furnace**, the heating elements come on in sequence as long as the temperature at the thermostat continues to drop

once the temperature starts to rise, no more elements are turned on

when there is minimal demand for heat, the furnace may be utilizing only one or two heating elements

0361

Fan limit switch

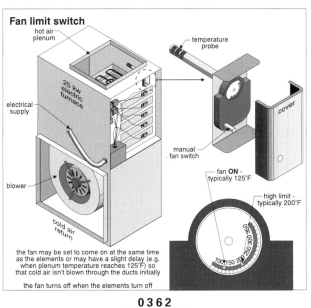

hot air plenum

temperature probe

25 kw electric furnace

electrical supply

cover

manual fan switch

blower

fan **ON** - typically 125°F

high limit - typically 200°F

cold air return

100 150 200 250 300 350

the fan may be set to come on at the same time as the elements or may have a slight delay (e.g. when plenum temperature reaches 125°F) so that cold air isn't blown through the ducts initially

the fan turns off when the elements turn off

0362

Plenum heater

return

hot air plenum

wood furnace

plenum heater

heating elements

supply

electrical supply

0363

Ampmeter testing of electric furnace

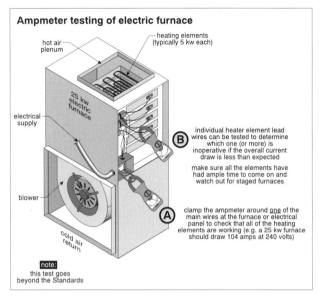

hot air plenum

heating elements (typically 5 kw each)

25 kw electric furnace

electrical supply

blower

cold air return

individual heater element lead wires can be tested to determine which one (or more) is inoperative if the overall current draw is less than expected

make sure all the elements have had ample time to come on and watch out for staged furnaces

clamp the ampmeter around <u>one</u> of the main wires at the furnace or electrical panel to check that all of the heating elements are working (e.g. a 25 kw furnace should draw 104 amps at 240 volts)

note:
this test goes beyond the Standards

0364

Excess temperature rise

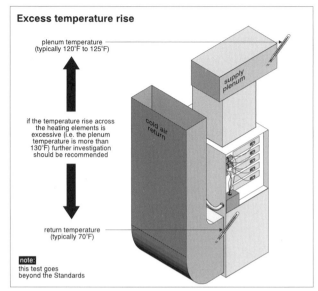

plenum temperature (typically 120°F to 125°F)

supply plenum

cold air return

if the temperature rise across the heating elements is excessive (i.e. the plenum temperature is more than 130°F) further investigation should be recommended

return temperature (typically 70°F)

note:
this test goes beyond the Standards

0365

Electric boiler

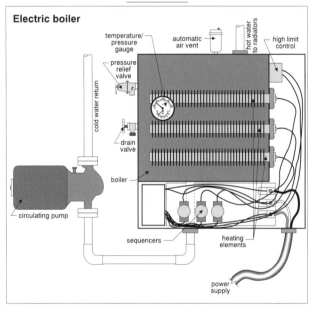

temperature/ pressure gauge

automatic air vent

hot water to radiators

high limit control

pressure relief valve

cold water return

drain valve

boiler

circulating pump

sequencers

heating elements

power supply

0366

Electric radiant heat - ceilings

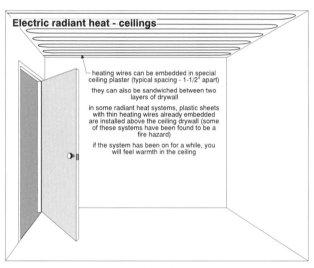

heating wires can be embedded in special ceiling plaster (typical spacing - 1-1/2" apart)

they can also be sandwiched between two layers of drywall

in some radiant heat systems, plastic sheets with thin heating wires already embedded are installed above the ceiling drywall (some of these systems have been found to be a fire hazard)

if the system has been on for a while, you will feel warmth in the ceiling

0367

Electric radiant heat - floors

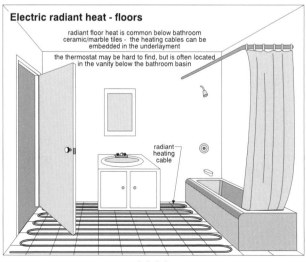

radiant floor heat is common below bathroom ceramic/marble tiles - the heating cables can be embedded in the underlayment

the thermostat may be hard to find, but is often located in the vanity below the bathroom basin

radiant heating cable

0368

Wall furnace

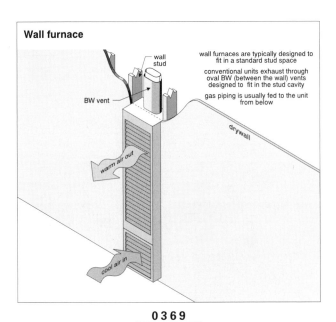

wall stud

BW vent

drywall

warm air out

cool air in

wall furnaces are typically designed to fit in a standard stud space

conventional units exhaust through oval BW (between the wall) vents designed to fit in the stud cavity

gas piping is usually fed to the unit from below

0369

Minimum wall furnace clearances

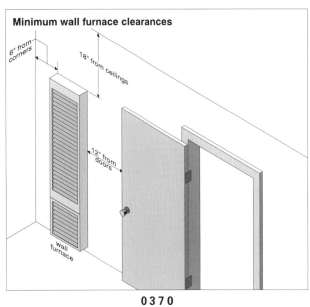

6" from corners

18" from ceilings

12" from doors

wall furnace

0370

Wall furnace venting

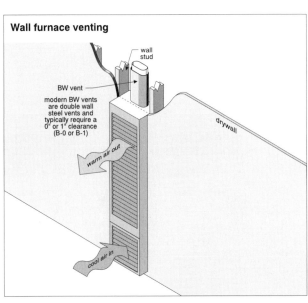

wall stud

BW vent

drywall

modern BW vents are double wall steel vents and typically require a 0" or 1" clearance (B-0 or B-1)

warm air out

cool air in

0371

Firestop spacers

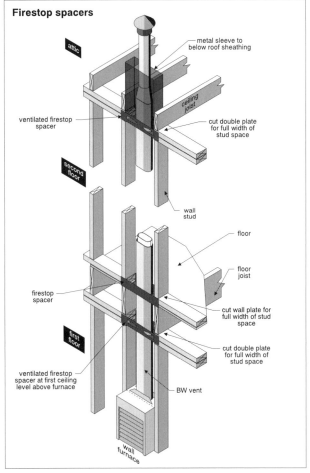

attic

metal sleeve to below roof sheathing

ceiling joist

ventilated firestop spacer

cut double plate for full width of stud space

second floor

wall stud

floor

firestop spacer

floor joist

cut wall plate for full width of stud space

cut double plate for full width of stud space

first floor

ventilated firestop spacer at first ceiling level above furnace

BW vent

wall furnace

0372

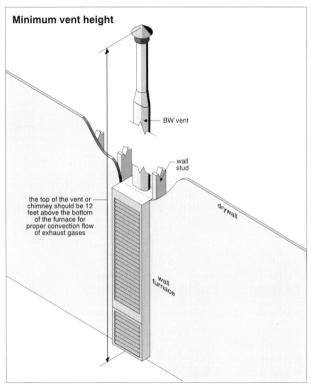

Minimum vent height

BW vent

wall stud

drywall

the top of the vent or chimney should be 12 feet above the bottom of the furnace for proper convection flow of exhaust gases

wall furnace

0373

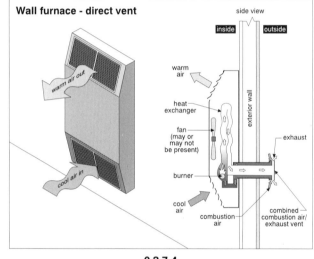

Wall furnace - direct vent

side view

inside   outside

warm air out

cool air in

warm air

heat exchanger

fan (may or may not be present)

burner

cool air

combustion air

exterior wall

exhaust

combined combustion air/ exhaust vent

0374

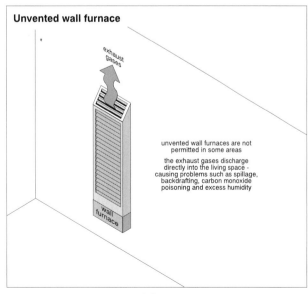

Unvented wall furnace

exhaust gases

unvented wall furnaces are not permitted in some areas

the exhaust gases discharge directly into the living space - causing problems such as spillage, backdrafting, carbon monoxide poisoning and excess humidity

wall furnace

0375

Minimum floor furnace clearances

60" below any structure (bulkhead, stairway)

6" from walls and corners

24"

6"

24" from two adjoining sides

24"

12" from doors, drapes, etc.

0376

Inspection cap missing or damaged

make sure that the inspection cap is in place and intact

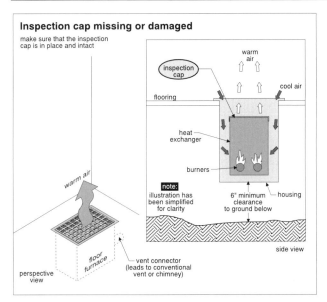

warm air

inspection cap

cool air

flooring

heat exchanger

burners

note: illustration has been simplified for clarity

6" minimum clearance to ground below

housing

side view

warm air

floor furnace

perspective view

vent connector (leads to conventional vent or chimney)

0377

Room heaters - combustible clearances

12" (varies) from sides and back

36" from top

vent connector

24" from front

18" (varies) from sides and back

36" from top

vent connector

36" from front

circulating type heater

note: combustible clearances can vary significantly from one jurisdiction to the next - check with your local authorities

radiant heater

0378

Unvented space heater

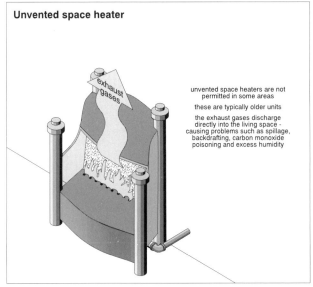

exhaust gases

unvented space heaters are not permitted in some areas

these are typically older units

the exhaust gases discharge directly into the living space - causing problems such as spillage, backdrafting, carbon monoxide poisoning and excess humidity

0379

Fireplace damper not fixed open

chimney

damper

if a room heater is installed in a fireplace, it should not be possible to close the fireplace damper

exhaust

room heater

cross section through chimney

floor joist

0380

Gas fireplace location restrictions

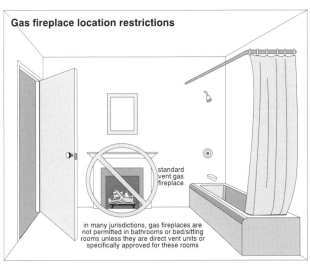

standard vent gas fireplace

in many jurisdictions, gas fireplaces are not permitted in bathrooms or bed/sitting rooms unless they are direct vent units or specifically approved for these rooms

0381

Gas fireplace - direct vent

inside

outside

exterior wall

combustion air inlet

flue gas exhaust

heated room air

radiant heat

combined combustion air/ exhaust vent

cross section

cool air from room

0382

PART 2

AIR CONDITIONING

AIR CONDITIONING

EVAPORATOR FAN (INDOOR FAN)

DUCT SYSTEM

THERMOSTATS

LIFE EXPECTANCY

EVAPORATIVE COOLERS

WHOLE HOUSE FANS

HEAT PUMPS

HEAT PUMPS IN THEORY

HEAT PUMPS IN PRACTICE

Moving heat from the inside to the outside

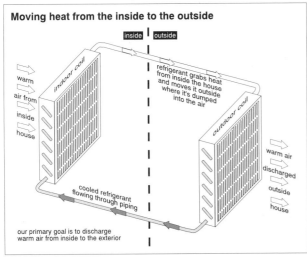

inside | outside

warm air from inside house

indoor coil

refrigerant grabs heat from inside the house and moves it outside where it's dumped into the air

outdoor coil

warm air discharged outside house

cooled refrigerant flowing through piping

our primary goal is to discharge warm air from inside to the exterior

0383

Heat transfer at the inside coil

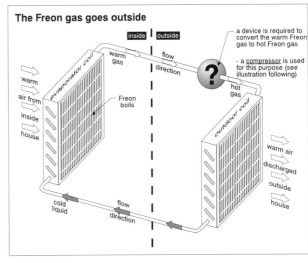

warm air

warmer

Freon flow direction

cooled air

Freon flow

aluminum fins

cold

Freon flow direction

in the coil, heat is transferred through aluminum fins attached to copper tubing that carries the refrigerant (Freon)

the fins are close together for maximum efficiency but this also makes it easy to clog the coil with dust and dirt

if we put cold Freon into the coil it will attract heat from the air passing around the fins

0384

Evaporator coil - collecting hot air inside the house

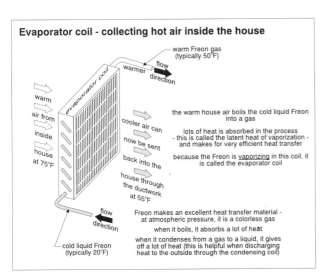

warm Freon gas (typically 50°F)

warmer

flow direction

warm air from inside house at 75°F

evaporator coil

cooler air can

now be sent

back into the

house through

the ductwork

at 55°F

flow direction

cold liquid Freon (typically 20°F)

the warm house air boils the cold liquid Freon into a gas

lots of heat is absorbed in the process - this is called the latent heat of vaporization - and makes for very efficient heat transfer

because the Freon is vaporizing in this coil, it is called the evaporator coil

Freon makes an excellent heat transfer material - at atmospheric pressure, it is a colorless gas

when it boils, it absorbs a lot of heat

when it condenses from a gas to a liquid, it gives off a lot of heat (this is helpful when discharging heat to the outside through the condensing coil)

0385

The Freon gas goes outside

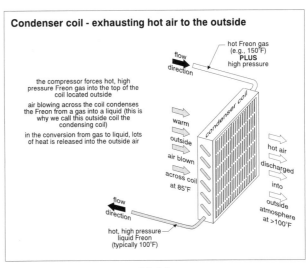

inside | outside

warm air from inside house

evaporator coil

warm gas

flow direction

Freon boils

a device is required to convert the warm Freon gas to hot Freon gas

- a compressor is used for this purpose (see illustration following)

hot gas

outdoor coil

warm air discharged outside house

cold liquid

flow direction

0386

Compressors - heating up Freon gas

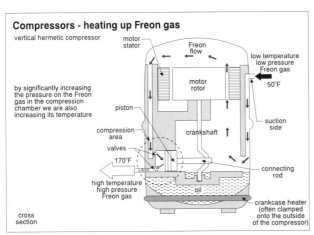

vertical hermetic compressor

motor stator

Freon flow

low temperature low pressure Freon gas

motor rotor

50°F

by significantly increasing the pressure on the Freon gas in the compression chamber we are also increasing its temperature

piston

compression area

valves

170°F

high temperature high pressure Freon gas

crankshaft

suction side

connecting rod

oil

crankcase heater (often clamped onto the outside of the compressor)

cross section

0387

Condenser coil - exhausting hot air to the outside

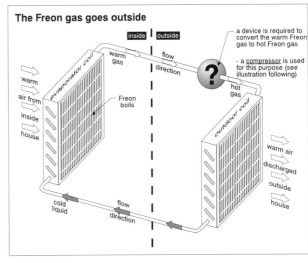

hot Freon gas (e.g., 150°F) PLUS high pressure

flow direction

the compressor forces hot, high pressure Freon gas into the top of the coil located outside

air blowing across the coil condenses the Freon from a gas into a liquid (this is why we call this outside coil the condensing coil)

in the conversion from gas to liquid, lots of heat is released into the outside air

condenser coil

warm outside air blown across coil at 85°F

hot air discharged into outside atmosphere at >100°F

flow direction

hot, high pressure liquid Freon (typically 100°F)

0388

Hot liquid back to house

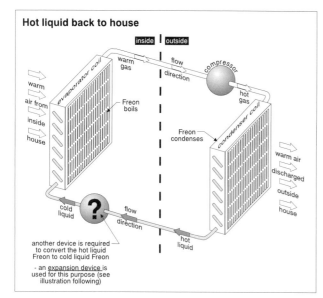

inside | outside

warm gas

flow direction

compressor

hot gas

evaporator coil

warm air from inside house

Freon boils

Freon condenses

condenser coil

warm air discharged outside house

cold liquid

?

flow direction

hot liquid

another device is required to convert the hot liquid Freon to cold liquid Freon

- an <u>expansion device</u> is used for this purpose (see illustration following)

0389

Expansion devices - cooling hot liquid Freon

the capillary tube is a bottleneck designed to restrict the flow of liquid Freon

at the discharge point of the bottleneck, the Freon is at a much lower pressure

when the pressure is reduced the temperature also goes down

the Freon coming out of the tube may be about 20°F and is ready to go into the evaporator coil to collect more heat

the expansion device is typically just upstream of the evaporator coil

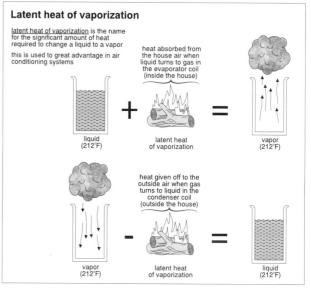

low temperature low pressure Freon liquid

high temperature high pressure Freon liquid

restriction

20°F

Freon flow

Freon flow

100°F

metering (or expansion) device

0390

Air conditioning - schematic of system

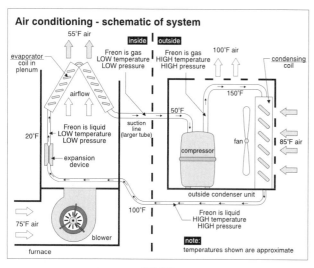

55°F air

inside | outside

Freon is gas LOW temperature LOW pressure

Freon is gas HIGH temperature HIGH pressure

100°F air

condensing coil

evaporator coil in plenum

airflow

150°F

50°F

20°F

Freon is liquid LOW temperature LOW pressure

suction line (larger tube)

fan

85°F air

expansion device

compressor

outside condenser unit

75°F air

blower

100°F

Freon is liquid HIGH temperature HIGH pressure

furnace

note: temperatures shown are approximate

0391

Latent heat of vaporization

<u>latent heat of vaporization</u> is the name for the significant amount of heat required to change a liquid to a vapor

this is used to great advantage in air conditioning systems

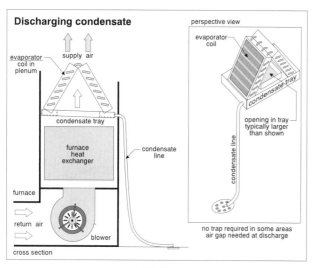

liquid (212°F)

latent heat of vaporization

vapor (212°F)

heat absorbed from the house air when liquid turns to gas in the evaporator coil (inside the house)

vapor (212°F)

latent heat of vaporization

liquid (212°F)

heat given off to the outside air when gas turns to liquid in the condenser coil (outside the house)

0392

Discharging condensate

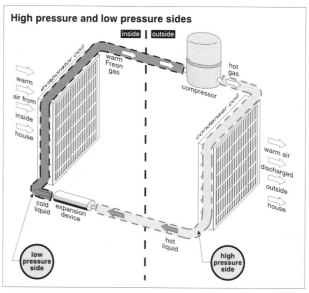

perspective view

evaporator coil in plenum

supply air

evaporator coil

condensate tray

furnace heat exchanger

condensate line

condensate tray

opening in tray typically larger than shown

condensate line

furnace

return air

blower

cross section

no trap required in some areas air gap needed at discharge

0393

High pressure and low pressure sides

inside | outside

evaporator coil

warm Freon gas

compressor

hot gas

warm air from inside house

condenser coil

warm air discharged outside house

cold liquid

expansion device

hot liquid

low pressure side

high pressure side

0394

Inspecting the condenser unit

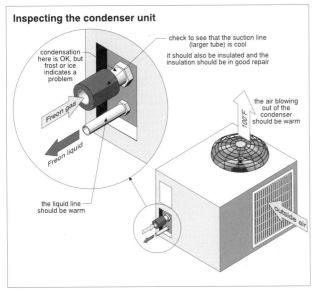

check to see that the suction line (larger tube) is cool

it should also be insulated and the insulation should be in good repair

condensation here is OK, but frost or ice indicates a problem

Freon gas

Freon liquid

the air blowing out of the condenser should be warm

100°F

the liquid line should be warm

outside air

0395

Air conditioning - schematic of system

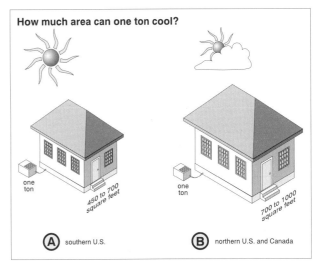

55°F air

inside | outside

100°F air

evaporator coil in plenum

Freon is gas LOW temperature LOW pressure

Freon is gas HIGH temperature HIGH pressure

condensing coil

airflow

150°F

20°F

Freon is liquid LOW temperature LOW pressure

suction line (larger tube)

50°F

expansion device

compressor

fan

85°F air

outside condenser unit

75°F air

100°F

Freon is liquid HIGH temperature HIGH pressure

blower

furnace

note:
temperatures shown are approximate

0396

One ton of cooling

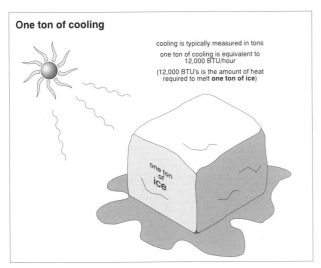

cooling is typically measured in tons

one ton of cooling is equivalent to 12,000 BTU/hour

(12,000 BTU's is the amount of heat required to melt **one ton of ice**)

one ton of ice

0397

How much area can one ton cool?

one ton

450 to 700 square feet

one ton

700 to 1000 square feet

(A) southern U.S.

(B) northern U.S. and Canada

0398

Larger ducts are required for air conditioning

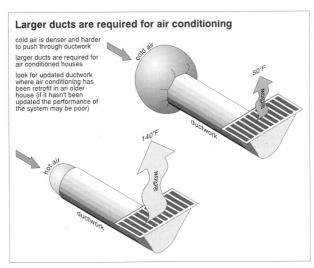

cold air is denser and harder to push through ductwork

larger ducts are required for air conditioned houses

look for updated ductwork where air conditioning has been retrofit in an older house (if it hasn't been updated the performance of the system may be poor)

cold air

50°F

airflow

ductwork

140°F

hot air

airflow

ductwork

0399

Bigger is not better

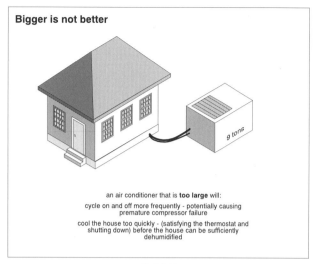

an air conditioner that is **too large** will:

cycle on and off more frequently - potentially causing premature compressor failure

cool the house too quickly - (satisfying the thermostat and shutting down) before the house can be sufficiently dehumidified

0400

Guessing the size

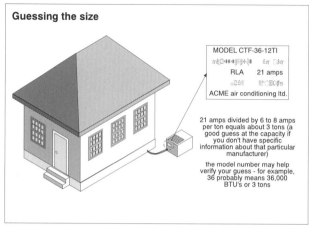

MODEL CTF-36-12TI

RLA   21 amps

ACME air conditioning ltd.

21 amps divided by 6 to 8 amps per ton equals about 3 tons (a good guess at the capacity if you don't have specific information about that particular manufacturer)

the model number may help verify your guess - for example, 36 probably means 36,000 BTU's or 3 tons

0401

Measure temperature drop across inside coil

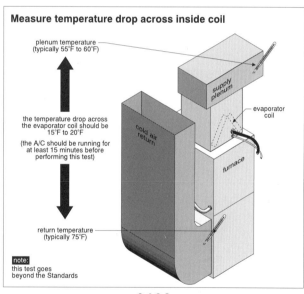

plenum temperature (typically 55°F to 60°F)

supply plenum

evaporator coil

cold air return

furnace

the temperature drop across the evaporator coil should be 15°F to 20°F

(the A/C should be running for at least 15 minutes before performing this test)

return temperature (typically 75°F)

note: this test goes beyond the Standards

0402

Don't test when it's cold

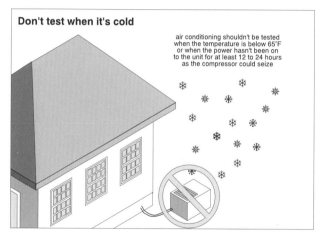

air conditioning shouldn't be tested when the temperature is below 65°F or when the power hasn't been on to the unit for at least 12 to 24 hours as the compressor could seize

0403

Wait before restarting a compressor

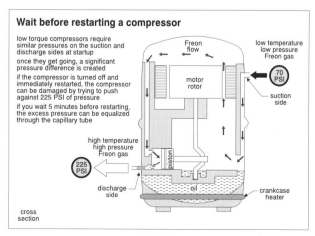

low torque compressors require similar pressures on the suction and discharge sides at startup

once they get going, a significant pressure difference is created

if the compressor is turned off and immediately restarted, the compressor can be damaged by trying to push against 225 PSI of pressure

if you wait 5 minutes before restarting, the excess pressure can be equalized through the capillary tube

Freon flow

low temperature low pressure Freon gas

70 PSI

suction side

motor rotor

high temperature high pressure Freon gas

225 PSI

piston

discharge side

oil

crankcase heater

cross section

0404

Slugging

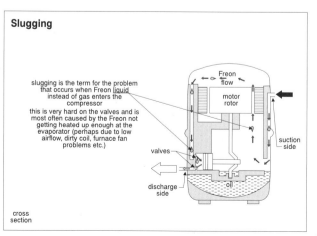

slugging is the term for the problem that occurs when Freon liquid instead of gas enters the compressor

this is very hard on the valves and is most often caused by the Freon not getting heated up enough at the evaporator (perhaps due to low airflow, dirty coil, furnace fan problems etc.)

Freon flow

motor rotor

suction side

valves

discharge side

oil

cross section

0405

Condensing unit out of level

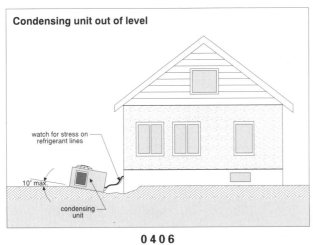

watch for stress on refrigerant lines

10° max.

condensing unit

0406

Excess electric current draw (beyond the Standards)

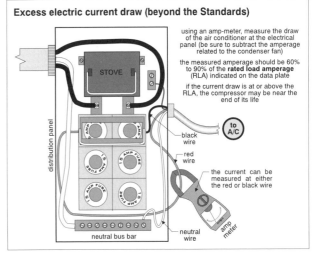

using an amp-meter, measure the draw of the air conditioner at the electrical panel (be sure to subtract the amperage related to the condenser fan)

the measured amperage should be 60% to 90% of the **rated load amperage** (RLA) indicated on the data plate

if the current draw is at or above the RLA, the compressor may be near the end of its life

STOVE

distribution panel

black wire

red wire

to A/C

the current can be measured at either the red or black wire

neutral bus bar

neutral wire

amp meter

0407

Missing electrical shut-off

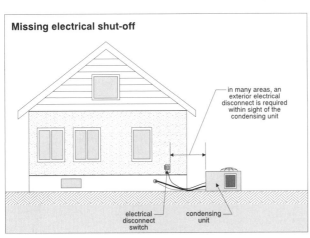

in many areas, an exterior electrical disconnect is required within sight of the condensing unit

electrical disconnect switch

condensing unit

0408

Water heater exhaust vent too close to condenser

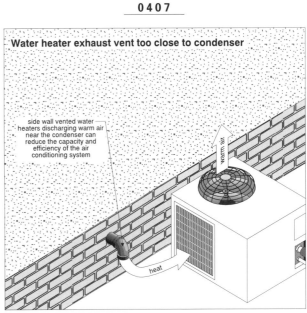

side wall vented water heaters discharging warm air near the condenser can reduce the capacity and efficiency of the air conditioning system

warm air

heat

0409

Dryer vent too close to condenser

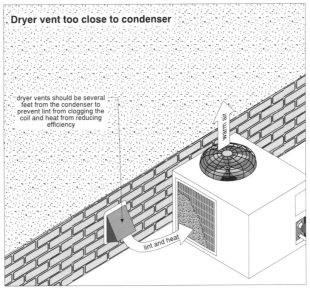

dryer vents should be several feet from the condenser to prevent lint from clogging the coil and heat from reducing efficiency

warm air

lint and heat

0410

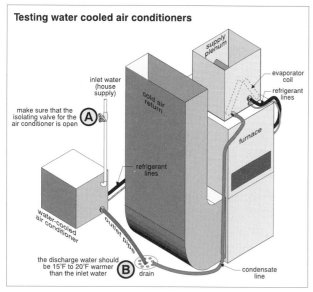

Testing water cooled air conditioners

inlet water (house supply)

make sure that the isolating valve for the air conditioner is open

(A)

supply plenum

cold air return

evaporator coil

refrigerant lines

furnace

refrigerant lines

water-cooled air conditioner

outlet pipe

the discharge water should be 15°F to 20°F warmer than the inlet water

(B) drain

condensate line

0 4 1 1

Missing backflow preventer

in some areas, an anti-siphon device such as a backflow preventer is required to prevent possible contamination of the drinking water in the event of a drop in the house water pressure

shut-off valve for A/C water supply

supply plenum

cold air return

evaporator coil

refrigerant lines

furnace

refrigerant lines

water-cooled air conditioner

outlet pipe

drain

condensate line

0 4 1 2

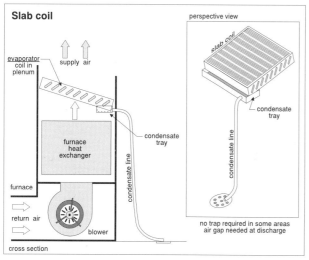

Slab coil

perspective view

evaporator coil in plenum

supply air

slab coil

condensate tray

furnace heat exchanger

condensate line

condensate tray

condensate line

furnace

return air

blower

no trap required in some areas air gap needed at discharge

cross section

0 4 1 3

Attic evaporator coil

attic evaporator coil

return ductwork

airflow

primary condensate line (condensate tray is typically built into the unit)

plumbing stack

trap

auxiliary condensate tray

supply ductwork

auxiliary condensate line

0 4 1 4

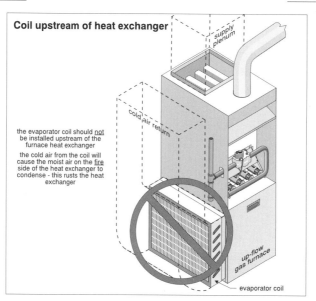

Coil upstream of heat exchanger

supply plenum

the evaporator coil should not be installed upstream of the furnace heat exchanger

the cold air from the coil will cause the moist air on the fire side of the heat exchanger to condense - this rusts the heat exchanger

cold air return

up-flow gas furnace

evaporator coil

0 4 1 5

Evaporator coil - inspection procedures

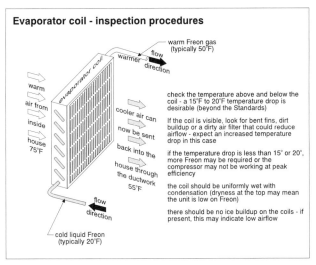

warm Freon gas (typically 50°F)

warmer flow direction

warm air from inside house 75°F

evaporator coil

cooler air can now be sent back into the house through the ductwork 55°F

check the temperature above and below the coil - a 15°F to 20°F temperature drop is desirable (beyond the Standards)

If the coil is visible, look for bent fins, dirt buildup or a dirty air filter that could reduce airflow - expect an increased temperature drop in this case

if the temperature drop is less than 15° or 20°, more Freon may be required or the compressor may not be working at peak efficiency

the coil should be uniformly wet with condensation (dryness at the top may mean the unit is low on Freon)

there should be no ice buildup on the coils - if present, this may indicate low airflow

flow direction

cold liquid Freon (typically 20°F)

0 4 1 6

Thermostatic expansion valve

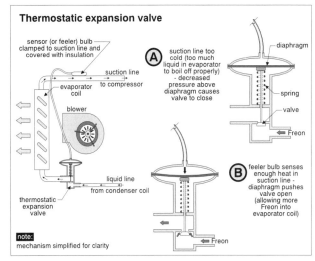

sensor (or feeler) bulb clamped to suction line and covered with insulation

suction line to compressor

evaporator coil

blower

liquid line from condenser coil

thermostatic expansion valve

Ⓐ suction line too cold (too much liquid in evaporator to boil off properly) - decreased pressure above diaphragm causes valve to close

diaphragm

spring

valve

Freon

Ⓑ feeler bulb senses enough heat in suction line - diaphragm pushes valve open (allowing more Freon into evaporator coil)

Freon

note: mechanism simplified for clarity

0 4 1 7

Capillary tube defects

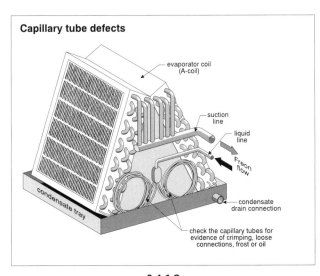

evaporator coil (A-coil)

suction line

liquid line

Freon flow

condensate tray

condensate drain connection

check the capillary tubes for evidence of crimping, loose connections, frost or oil

0 4 1 8

Leaking condensate tray

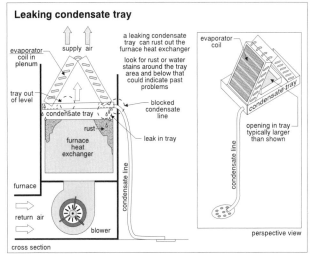

supply air

a leaking condensate tray can rust out the furnace heat exchanger

look for rust or water stains around the tray area and below that could indicate past problems

evaporator coil in plenum

tray out of level

condensate tray

rust

furnace heat exchanger

furnace

return air

blower

blocked condensate line

leak in tray

condensate line

evaporator coil

condensate tray

opening in tray typically larger than shown

condensate line

perspective view

cross section

0 4 1 9

Auxiliary condensate line from attic evaporator coil

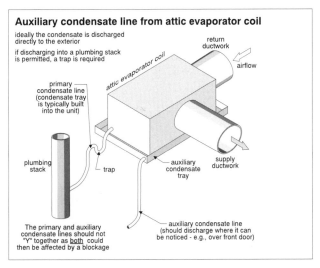

ideally the condensate is discharged directly to the exterior

if discharging into a plumbing stack is permitted, a trap is required

return ductwork

airflow

attic evaporator coil

primary condensate line (condensate tray is typically built into the unit)

supply ductwork

plumbing stack

trap

auxiliary condensate tray

auxiliary condensate line (should discharge where it can be noticed - e.g., over front door)

The primary and auxiliary condensate lines should not "Y" together as <u>both</u> could then be affected by a blockage

0 4 2 0

Trap required in condensate line

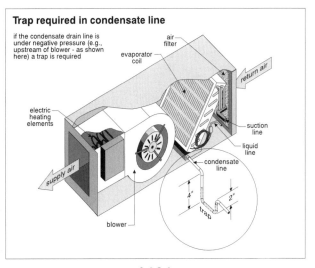

if the condensate drain line is under negative pressure (e.g., upstream of blower - as shown here) a trap is required

air filter

evaporator coil

return air

electric heating elements

supply air

blower

suction line

liquid line

condensate line

4"

2"

trap

0 4 2 1

Condensate pump

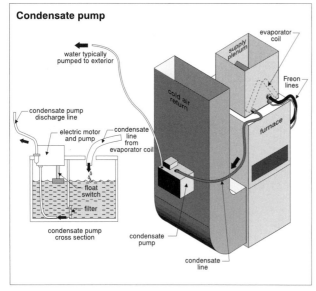

water typically pumped to exterior

condensate pump discharge line

electric motor and pump

condensate line from evaporator coil

cold air return

evaporator coil

supply plenum

Freon lines

furnace

float switch

filter

condensate pump cross section

condensate pump

condensate line

condensate line

0422

Refrigerant lines

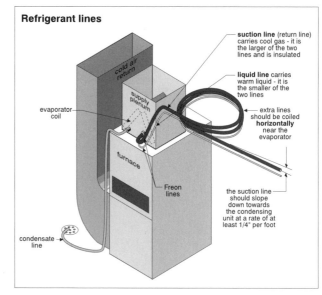

suction line (return line) carries cool gas - it is the larger of the two lines and is insulated

liquid line carries warm liquid - it is the smaller of the two lines

extra lines should be coiled **horizontally** near the evaporator

cold air return

supply plenum

evaporator coil

furnace

Freon lines

condensate line

the suction line should slope down towards the condensing unit at a rate of at least 1/4" per foot

0423

Filter/dryer

in some installations, (especially larger units) you will find a filter/dryer in the liquid line

it can be near the condenser, near the expansion device or in the condenser cabinet

filter/dryers are often retrofitted after compressor replacement to remove any remaining contaminants

a filter/dryer typically contains a cotton filter and silica gel as a drying agent

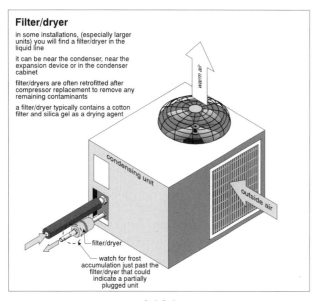

warm air

condensing unit

outside air

filter/dryer

watch for frost accumulation just past the filter/dryer that could indicate a partially plugged unit

0424

Sight glass

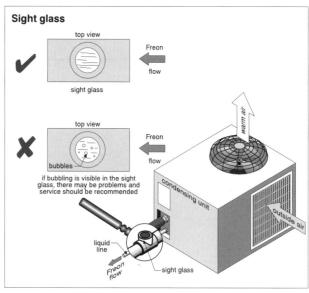

top view

Freon flow

sight glass

top view

Freon flow

bubbles

if bubbling is visible in the sight glass, there may be problems and service should be recommended

warm air

condensing unit

outside air

liquid line

Freon flow

sight glass

0425

Leaking refrigerant lines - vulnerable areas

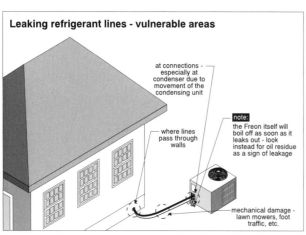

at connections - especially at condenser due to movement of the condensing unit

where lines pass through walls

note:
the Freon itself will boil off as soon as it leaks out - look instead for oil residue as a sign of leakage

mechanical damage - lawn mowers, foot traffic, etc.

0426

Condenser fan

the fans in most modern condenser units rotate horizontally and discharge air out the top (air is brought in through the sides) but, they can also operate diagonally and vertically depending on the manufacturer

excess vibration or bearing noise may indicate that bearing failure is a potential problem

the fan blades should be turning very fast when the unit is in operation (if they aren't - service is required)

to prevent rusting of the fan and motor, condensing units should be covered in the winter according to some experts

100°F

data plate

condenser unit

outside air

Freon lines

0427

Condenser coil location requirements

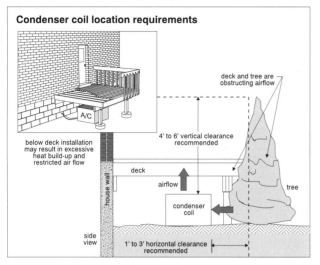

below deck installation may result in excessive heat build-up and restricted air flow

house wall

A/C

deck and tree are obstructing airflow

4' to 6' vertical clearance recommended

deck

airflow

condenser coil

tree

side view

1' to 3' horizontal clearance recommended

0428

Evaporator fan

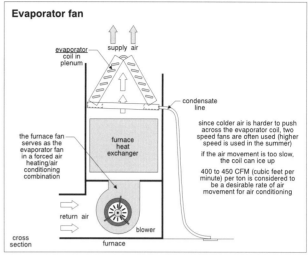

evaporator coil in plenum

supply air

condensate line

the furnace fan serves as the evaporator fan in a forced air heating/air conditioning combination

furnace heat exchanger

since colder air is harder to push across the evaporator coil, two speed fans are often used (higher speed is used in the summer)

if the air movement is too slow, the coil can ice up

400 to 450 CFM (cubic feet per minute) per ton is considered to be a desirable rate of air movement for air conditioning

return air

blower

cross section

furnace

0429

Belt or pulley adjustment

a loose or misaligned fan belt can reduce the amount of air flowing past the evaporator coil and degrade system performance

-check belt for cracks or other wear
-check belt tension (see below)
-check for excess vibration
-check for overheating at the motor

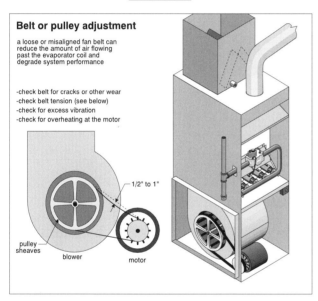

1/2" to 1"

pulley sheaves

blower

motor

0430

Rule of thumb for ductwork adequacy

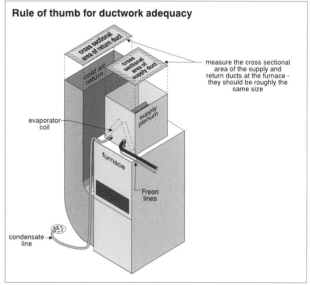

cross sectional area of return duct

cross sectional area of supply duct

cold air return

measure the cross sectional area of the supply and return ducts at the furnace - they should be roughly the same size

evaporator coil

supply plenum

furnace

Freon lines

condensate line

0431

Flow of cooled air - older-style ductwork

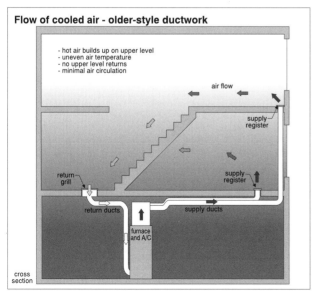

- hot air builds up on upper level
- uneven air temperature
- no upper level returns
- minimal air circulation

air flow

supply register

return grill

supply register

return ducts

supply ducts

furnace and A/C

cross section

0432

Flow of cooled air - modern ductwork

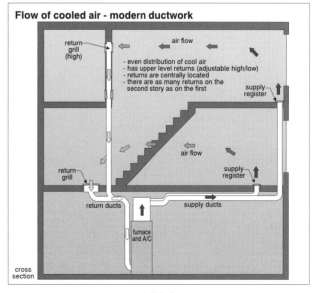

return grill (high)

air flow

- even distribution of cool air
- has upper level returns (adjustable high/low)
- returns are centrally located
- there are as many returns on the second story as on the first

supply register

return grill

air flow

supply register

return ducts

supply ducts

furnace and A/C

cross section

0433

High and low returns

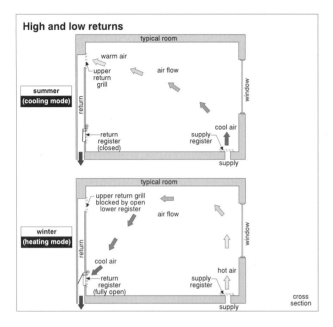

typical room

warm air

upper return grill

air flow

summer (cooling mode)

return

cool air

return register (closed)

supply register

cool air

supply

window

typical room

upper return grill blocked by open lower register

air flow

winter (heating mode)

return

cool air

return register (fully open)

supply register

hot air

supply

window

cross section

__0 4 3 4__

Testing cold air returns

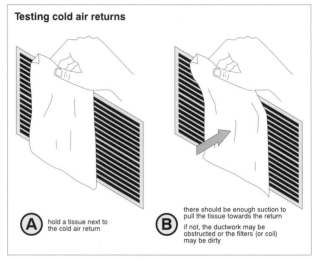

A hold a tissue next to the cold air return

B there should be enough suction to pull the tissue towards the return

if not, the ductwork may be obstructed or the filters (or coil) may be dirty

__0 4 3 5__

Ducts in concrete floor slabs

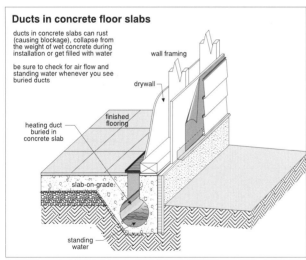

ducts in concrete slabs can rust (causing blockage), collapse from the weight of wet concrete during installation or get filled with water

be sure to check for air flow and standing water whenever you see buried ducts

wall framing

drywall

finished flooring

heating duct buried in concrete slab

slab-on-grade

standing water

__0 4 3 6__

Air return outside room

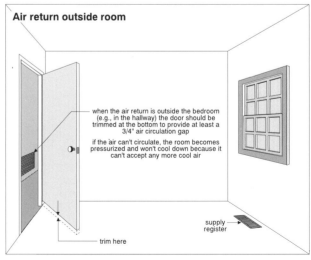

when the air return is outside the bedroom (e.g., in the hallway) the door should be trimmed at the bottom to provide at least a 3/4" air circulation gap

if the air can't circulate, the room becomes pressurized and won't cool down because it can't accept any more cool air

trim here

supply register

__0 4 3 7__

Close humidifier damper in summer

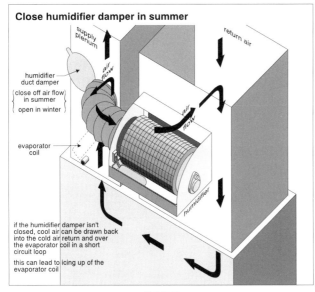

supply plenum

return air

air flow

humidifier duct damper
(close off air flow in summer)
open in winter

air flow

evaporator coil

humidifier

if the humidifier damper isn't closed, cool air can be drawn back into the cold air return and over the evaporator coil in a short circuit loop

this can lead to icing up of the evaporator coil

0438

Vapor barriers and air conditioning ductwork

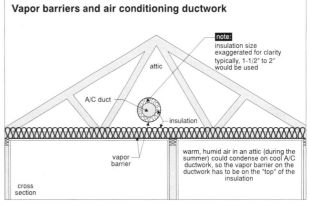

note: insulation size exaggerated for clarity typically, 1-1/2" to 2" would be used

attic

A/C duct

insulation

vapor barrier

warm, humid air in an attic (during the summer) could condense on cool A/C ductwork, so the vapor barrier on the ductwork has to be on the "top" of the insulation

cross section

0439

Thermostat - bi-metallic (mercury bulb)

the central hub of the bi-metallic coil is attached to the thermostat temperature setting dial - turning the dial down (counterclockwise) means that it will have to get even cooler before the mercury rolls down and closes the contacts

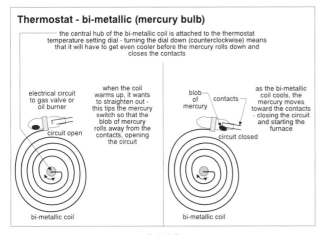

electrical circuit to gas valve or oil burner

when the coil warms up, it wants to straighten out - this tips the mercury switch so that the blob of mercury rolls away from the contacts, opening the circuit

circuit open

bi-metallic coil

blob of mercury

contacts

as the bi-metallic coil cools, the mercury moves toward the contacts - closing the circuit and starting the furnace

circuit closed

bi-metallic coil

0440

Thermostat - bi-metallic (snap-action)

the central hub of the bi-metallic coil is attached to the thermostat temperature setting dial - turning the dial down (counterclockwise) means that it will have to get even cooler before the magnet activates the circuit

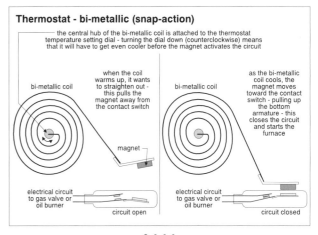

bi-metallic coil

when the coil warms up, it wants to straighten out - this pulls the magnet away from the contact switch

magnet

electrical circuit to gas valve or oil burner

circuit open

bi-metallic coil

as the bi-metallic coil cools, the magnet moves toward the contact switch - pulling up the bottom armature - this closes the circuit and starts the furnace

electrical circuit to gas valve or oil burner

circuit closed

0441

Poor location for thermostat

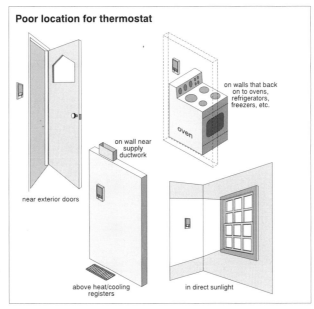

on walls that back on to ovens, refrigerators, freezers, etc.

oven

on wall near supply ductwork

near exterior doors

above heat/cooling registers

in direct sunlight

0442

Typical compressor life

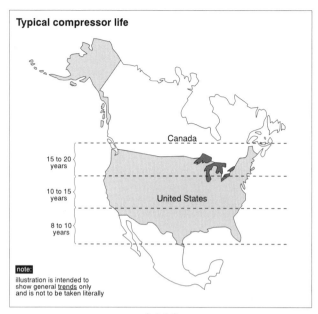

Canada

15 to 20
years

10 to 15
years

United States

8 to 10
years

note:
illustration is intended to
show general <u>trends</u> only
and is not to be taken literally

0 4 4 3

Three types of evaporative coolers

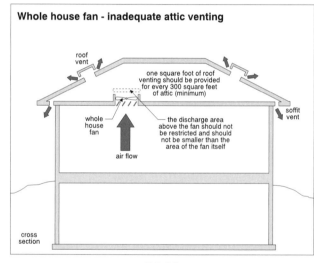

rotary

outdoor
air

drum
rotation
motor
water
reservoir
float
valve

blower

to
house

spray

outdoor
air

evaporative
pad

spray

water
slinger

water
reservoir

float
valve

blower

to
house

drip

evaporative
pad

outdoor
air

distribution
tube

recirculating
pump

water
reservoir

float
valve

blower

to
house

0 4 4 4

Whole house fan

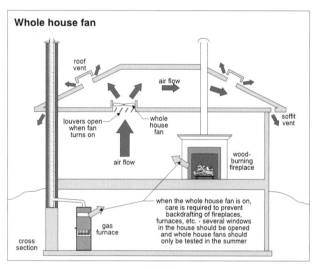

roof
vent

air flow

louvers open
when fan
turns on

whole
house
fan

air flow

wood-
burning
fireplace

soffit
vent

cross
section

gas
furnace

when the whole house fan is on,
care is required to prevent
backdrafting of fireplaces,
furnaces, etc. - several windows
in the house should be opened
and whole house fans should
only be tested in the summer

0 4 4 5

Whole house fan - inadequate attic venting

roof
vent

one square foot of roof
venting should be provided
for every 300 square feet
of attic (minimum)

whole
house
fan

air flow

the discharge area
above the fan should not
be restricted and should
not be smaller than the
area of the fan itself

soffit
vent

cross
section

0 4 4 6

Heat pump principles

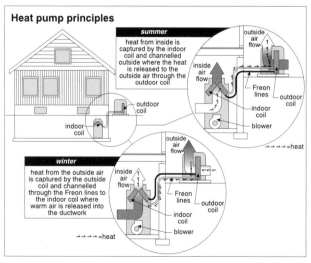

summer
heat from inside is
captured by the indoor
coil and channelled
outside where the heat
is released to the
outside air through the
outdoor coil

outside
air
flow

inside
air
flow

Freon
lines

indoor
coil

outdoor
coil

blower

outdoor
coil

indoor
coil

=heat

winter
heat from the outside air
is captured by the outside
coil and channelled
through the Freon lines to
the indoor coil where
warm air is released into
the ductwork

outside
air
flow

inside
air
flow

Freon
lines

indoor
coil

outdoor
coil

blower

=heat

0 4 4 7

Heat pump evaporator coil - grabbing heat from the outside air

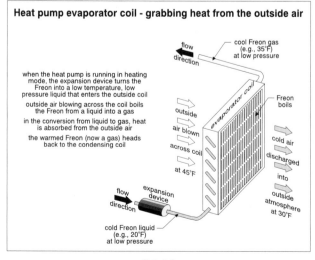

flow
direction

cool Freon gas
(e.g., 35°F)
at low pressure

when the heat pump is running in heating
mode, the expansion device turns the
Freon into a low temperature, low
pressure liquid that enters the outside coil

outside air blowing across the coil boils
the Freon from a liquid into a gas

in the conversion from liquid to gas, heat
is absorbed from the outside air

the warmed Freon (now a gas) heads
back to the condensing coil

outside
air blown

across coil

at 45°F

Freon
boils

cold air

discharged

into

outside
atmosphere
at 30°F

flow
direction

expansion
device

cold Freon liquid
(e.g., 20°F)
at low pressure

0 4 4 8

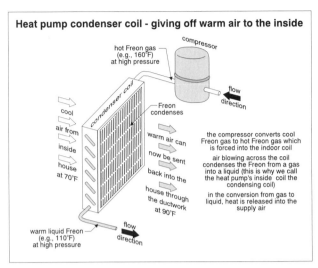

Heat pump condenser coil - giving off warm air to the inside

compressor

hot Freon gas (e.g., 160°F) at high pressure

flow direction

condenser coil

Freon condenses

cool air from inside house at 70°F

warm air can now be sent back into the house through the ductwork at 90°F

warm liquid Freon (e.g., 110°F) at high pressure

flow direction

the compressor converts cool Freon gas to hot Freon gas which is forced into the indoor coil

air blowing across the coil condenses the Freon from a gas into a liquid (this is why we call the heat pump's inside coil the condensing coil)

in the conversion from gas to liquid, heat is released into the supply air

0449

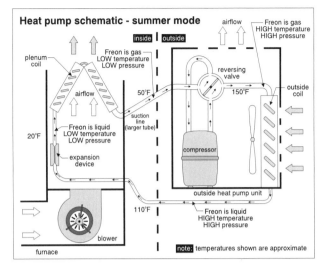

Heat pump schematic - summer mode

airflow

Freon is gas HIGH temperature HIGH pressure

inside | outside

plenum coil

Freon is gas LOW temperature LOW pressure

airflow

reversing valve

outside coil

20°F

50°F

150°F

Freon is liquid LOW temperature LOW pressure

suction line (larger tube)

expansion device

compressor

outside heat pump unit

110°F

Freon is liquid HIGH temperature HIGH pressure

blower

furnace

note: temperatures shown are approximate

0450

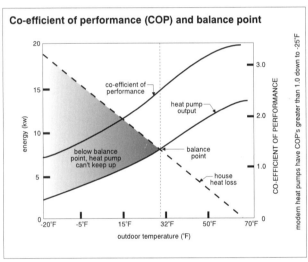

Co-efficient of performance (COP) and balance point

co-efficient of performance

heat pump output

balance point

below balance point, heat pump can't keep up

house heat loss

energy (kw)

outdoor temperature (°F)

CO-EFFICIENT OF PERFORMANCE

modern heat pumps have COP's greater than 1.0 down to -25°F

0451

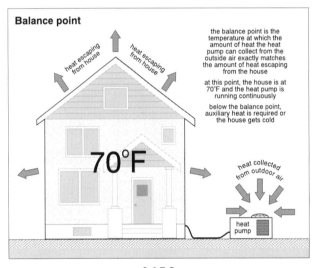

Balance point

heat escaping from house

heat escaping from house

70°F

heat collected from outdoor air

heat pump

the balance point is the temperature at which the amount of heat the heat pump can collect from the outside air exactly matches the amount of heat escaping from the house

at this point, the house is at 70°F and the heat pump is running continuously

below the balance point, auxiliary heat is required or the house gets cold

0452

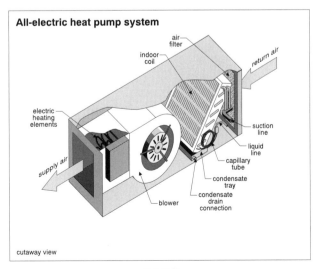

All-electric heat pump system

air filter

indoor coil

return air

electric heating elements

suction line

liquid line

capillary tube

condensate tray

condensate drain connection

supply air

blower

cutaway view

0453

Triple split system heat pump

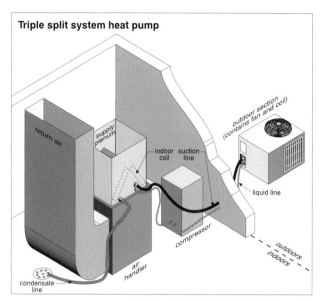

0454

Two expansion devices

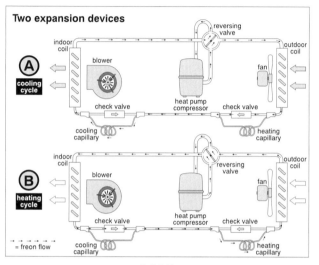

0455

Defrost cycle

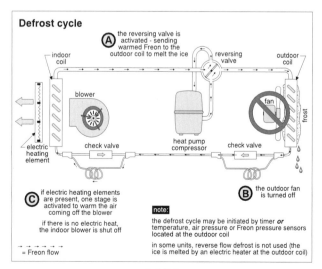

0456

Poor outdoor coil location

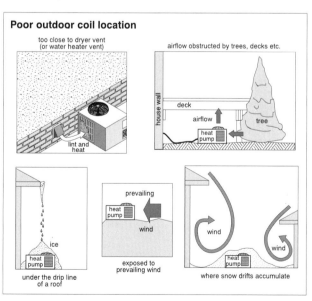

0457

Testing back-up heat

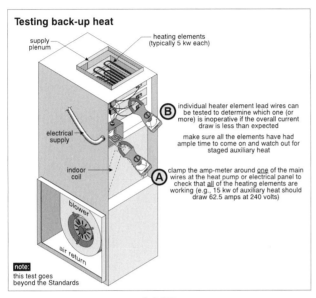

supply plenum

heating elements (typically 5 kw each)

electrical supply

indoor coil

blower

air return

B individual heater element lead wires can be tested to determine which one (or more) is inoperative if the overall current draw is less than expected

make sure all the elements have had ample time to come on and watch out for staged auxiliary heat

A clamp the amp-meter around <u>one</u> of the main wires at the heat pump or electrical panel to check that <u>all</u> of the heating elements are working (e.g., 15 kw of auxiliary heat should draw 62.5 amps at 240 volts)

note:
this test goes beyond the Standards

0458

Water source heat pump - open loop system (well-based)

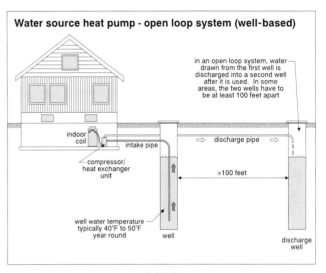

in an open loop system, water drawn from the first well is discharged into a second well after it is used. In some areas, the two wells have to be at least 100 feet apart

indoor coil

intake pipe

discharge pipe

compressor/ heat exchanger unit

>100 feet

well water temperature typically 40°F to 50°F year round

well

discharge well

0459

Ground source heat pump - horizontal closed loop

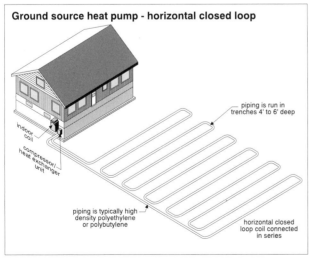

indoor coil

compressor/ heat exchanger unit

piping is run in trenches 4' to 6' deep

piping is typically high density polyethylene or polybutylene

horizontal closed loop coil connected in series

0460

Ground source heat pump - vertical closed loop

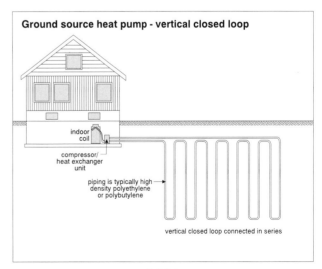

indoor coil

compressor/ heat exchanger unit

piping is typically high density polyethylene or polybutylene

vertical closed loop connected in series

0461

Bivalent schematic

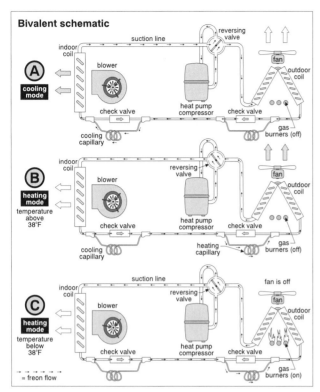

0462

With over 500 illustrations, *Structure, Roofing and the Exterior* shows you exactly what the finished job should look like. This book is an excellent tool for the new home buyer or inspector, showing you which pitfalls to avoid and what to look for when searching for potential problems in the structure, roof or exterior of your or your client's home.

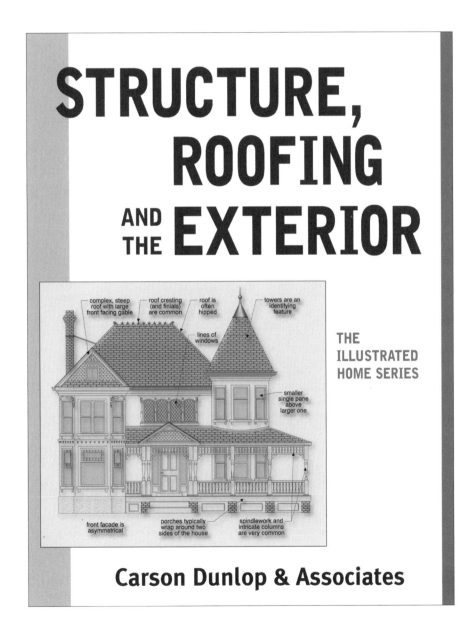

STRUCTURE, ROOFING AND THE EXTERIOR

THE ILLUSTRATED HOME SERIES

complex, steep roof with large front facing gable

roof cresting (and finials) are common

roof is often hipped

towers are an identifying feature

lines of windows

smaller single pane above larger one

front facade is asymmetrical

porches typically wrap around two sides of the house

spindlework and intricate columns are very common

Carson Dunlop & Associates

With more than 450 illustrations, *Electrical, Plumbing, Insulation and the Interior* shows you exactly what the finished job should look like. This book is an excellent tool for the new home buyer or inspector, showing you which pitfalls to avoid and what to look for when searching for potential problems with the electrical work, plumbing and insulation of your or your client's home.

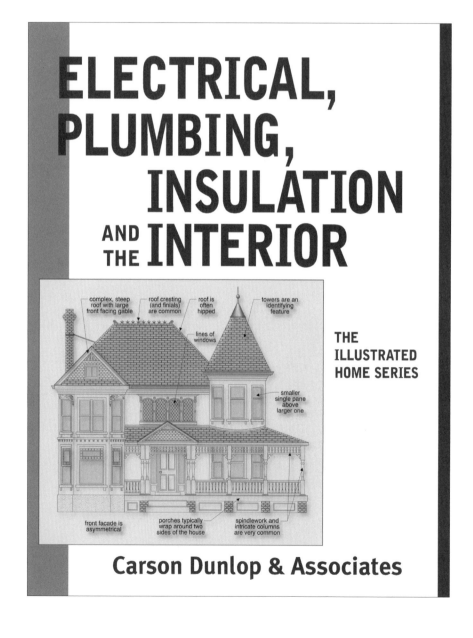